FICAN

Federal Interagency Committee on Aviation Noise

Assessment of Tools for Modeling Aircraft Noise in the National Parks

Gregg G. Fleming
Kenneth J. Plotkin
Christopher J. Roof
Bruce J. Ikelheimer
David A. Senzig

March 18, 2005

USDOT Research & Special Programs Administration
John A. Volpe National Transportation Systems Center
Environmental Measurement and Modeling Division
Acoustics Facility

 *Wyle Laboratories*

Table of Contents

Section   Page

Table of Contents i
List of Tables iii
List of Figures iv
List of Acronyms vii
1. Introduction 1
   1.1 Study Background and Introduction to the Models 1
      1.1.1 INM 2
      1.1.2 NMSim 3
   1.2 Objectives 4
   1.3 Organization of Document 4
2. Comparison of Model Capabilities 5
   2.1 Source Characterization 6
      2.1.1 Description 6
      2.1.2 Fleet Coverage 8
   2.2 Propagation 12
      2.2.1 Atmospheric Absorption 13
      2.2.2 Lateral Effects 14
      2.2.3 Terrain Shielding 16
   2.3 Contouring/Grid Development and Noise Metrics 16
   2.4 Simulation vs. Integrated Models 18
   2.5 Calculating Audibility 19
   2.6 Other Capabilities 19
3. Comparison of Model Calculations 23
   3.2 Grand Canyon Noise Model Validation Study (GCNP MVS) 27
      3.2.1 Grand Canyon MVS Sensitivities 30
4. Grand Canyon Noise Analysis 39
   4.1 Contributions by Category 39
      4.1.1 Aircraft Scenario 1 39
      4.1.2 Aircraft Scenario 2 41
      4.1.3 Aircraft Scenario 3 42
      4.1.4 Aircraft Scenario 4 43
      4.1.5 Aircraft Scenario 5 44
      4.1.6 Aircraft Scenario 6 44
      4.1.7 Aircraft Scenarios 7 to 10 45
   4.2 Aggregate Contributions 50
   4.3 Contour Sensitivities 51
   4.4 Margin of Safety (Contours with uncertainties) 52
5. Model Usability 55
   5.1 Data Requirements 55
   5.2 User Interface 56
   5.3 Outputs 56
   5.4 Implementation 57
   5.5 Other 59

Table of Contents (continued)

Section	Page
6. Summary of Findings and Recommended Improvements to the Models	61
7. References	63

Appendix A: FAA/NPS Letter to FICAN, Terms of Reference and Statement of Work... A-1

Appendix B: Summary of Enhancements Included in INM Version 6.2 for use in the National Parks B-1
    B.1 Summary of INM 6.2 Updates B-2
    B.2 Commercial Aircraft Noise/Performance Database B-2
    B.3 Noise Modeling for National Parks B-2
        B.3.1 New Noise Metrics B-3
        B.3.2 Using TAUD and DDOSE in INM B-3
        B.3.3. National Parks Noise Database Enhancements B-5
        B.3.4. Terrain Modeling - Line-of-Sight Blockage B-6
        B.3.5. Terrain Modeling – Additional Terrain Data Capability B-7
        B.3.6. Disabling Lateral Attenuation for Propeller Aircraft B-8
        B.3.7. Level Flyover NPD Curves B-8
    B.4. New MapInfo Interchange File Export Function B-9
    B.5. Database Modifications B-9
    B.6. Program Modifications B-13
    B.7. Reported Problems Fixed B-16

Appendix C: Summary of Commercial Jet Overflights in GCNP, August 31, 2003 .. C-1

Appendix D: Theoretical GCNP Jet Audibility Assessment D-1

Appendix E: GCNP Sound Level Measurements of High Altitude Jet Aircraft E-1

Appendix F: Statistical Definitions F-1

Appendix G: Development of Reference Noise Data for High Altitude Jets G-1

Appendix H: Audibility Calculations for National Parks H-1

List of Tables

Table		Page
1.	Comparison of Features in INM and NMSim	5
2.	Aircraft Represented in the INM NPD Database	9
3.	Comparison of Absorption Coefficients, SAE versus ISO	13
4.	Noise Metrics in INM	17
5.	Noise Metrics in NMSim	18
6.	%T_{Aud} Statistics	29
7.	Comparison of Terrain Data Elevations	31
8.	Summary of GCNP Model Sensitivities	52
9.	Summary of Lower Bounds to Contour Area Uncertainty GCNP Aircraft Scenario 11	53
10.	Comparison of Core Computational Run Time (minutes)	60
B-1.	One-Third Octave Band Characteristics	24
B-2.	Equivalent Auditory System Noise (EASN)	27
B-3.	INM 6.2 Lateral Attenuation Algorithm Update	29
B-4.	Spectral Class Assignments by Aircraft Type	29
C-1.	Comparison of ETMS and PDARS Data	7
D-1.	Aircraft Corrected Net Thrust (F_n/δ, pounds) as a Function of Altitude	2
D-2.	L_{ASmx} values at 1000 feet; INM	8
D-3.	L_{ASmx} values at 1000 feet; McAninch Model	8
D-4.	Difference in L_{ASmx} values at 1000 feet; INM minus McAninch Model	8
E-1.	Hancock Knoll Measurement Summary Statistics	6
E-2.	Swamp Point Measurement Summary Statistics	7
F-1.	Definitions of Statistical Measures	2

List of Figures

Figure		Page
1.	INM Development Timeline	2
2.	Example NPD Data for DeHaviland DHC-6 with Quiet Propellers	6
3.	Percentage of the Active Aviation Fleet Represented by the INM NPD Database	9
4.	Summary of Sound Level Differences due to Differing Atmospheric Absorption	14
5.	Lateral Directivity in INM for Jet Aircraft	15
6.	Approximation of Noise Level Time History by Simulation and Integrated Noise Models	21
7.	Comparison of INM and NMSim SEL NPD Data	24
8.	Comparison of INM and NMSim L_{ASmx} NPD Data	24
9.	Comparison of INM and NMSim Atmospheric Absorption Effects on Noise Data	25
10.	Comparison of INM and NMSim Lateral Effects	25
11.	Comparison of INM and NMSim Terrain Shielding	26
12.	Comparison of INM and NMSim Contouring	27
13.	INM 6.2 and NMSim Modeled vs. Measured %TAud	28
14.	Grand Canyon Noise Model Validation Study Figure 12	28
15.	$\%T_{Aud}$ Model Bias and CIs, Individual Hours	29
16.	$\%T_{Aud}$ Model Bias and CIs, Site Groups	30
17.	Comparison of INM and NMSim Time Audible With Terrain	32
18.	Comparison of INM and NMSim Time Audible with Flat Earth	33
19.	Comparison of INM and NMSim Time Audible with Flat Earth and Identical NPDs	34
20.	Single-Event Time Audible Comparisons	35
21.	Effect of Overlap on %TAUD NMSim Scheduler Results Compared with INM Compression Algorithm	37
22.	25 $\%T_{Aud,}$ Aircraft Scenario 1, INM	40
23.	25 $\%T_{Aud,}$ Aircraft Scenario 1, NMSim	40
24.	25 $\%T_{Aud,}$ Aircraft Scenario 2, INM	41
25.	25 $\%T_{Aud,}$ Aircraft Scenario 2, NMSim	41
26.	25 $\%T_{Aud,}$ Aircraft Scenario 3, INM	42
27.	25 $\%T_{Aud,}$ Aircraft Scenario 3, NMSim	42
28.	25 $\%T_{Aud,}$ Aircraft Scenario 4, INM	43
29.	25 $\%T_{Aud,}$ Aircraft Scenario 4, NMSim	43
30.	25 $\%T_{Aud,}$ Aircraft Scenario 6, INM	44
31.	25 $\%T_{Aud,}$ Aircraft Scenario 6, NMSim 25% TAud = 31.4% of Park	45
32.	25 $\%T_{Aud,}$ Aircraft Scenario 7, INM	46
33.	25 $\%T_{Aud,}$ Aircraft Scenario 7, NMSim	46
34.	25 $\%T_{Aud,}$ Aircraft Scenario 8, INM	47
35.	25 $\%T_{Aud,}$ Aircraft Scenario 8, NMSim	47
36.	25 $\%T_{Aud,}$ Aircraft Scenario 9, INM	48
37.	25 $\%T_{Aud,}$ Aircraft Scenario 9, NMSim	48
38.	25 $\%T_{Aud,}$ Aircraft Scenario 10, INM	49
39.	25 $\%T_{Aud,}$ Aircraft Scenario 10, NMSim –	49
40.	25 $\%T_{Aud,}$ Aircraft Scenario 11: Combination of Scenarios 1 to 5, INM	50

Table of Figures (continued)

<u>Figure</u> <u>Page</u>

41.	25 %T_{Aud}, Aircraft Scenario 11: Combination of Scenarios 1 to 5, NMSim	51
42.	Comparison of Measured Audibility in GCNP	53
B-1.	Line-of-Sight (LOS) Blockage Concept	6
B-2.	One-Third Octave Band Equivalent Auditory System Noise (EASN) Floor	26
B-3.	Long-Range Air-to-Ground Attenuation $\Lambda(\beta)$	28
C-1.	ORD Flights over GCNP	2
C-2.	DEN Flights over GCNP	3
C-3.	JFK Flights over GCNP	3
C-4.	LAS Flights over GCNP	4
C-5.	LAX Flights over GCNP	4
C-6.	PHX Flights over GCNP	5
C-8.	All Flights over GCNP	6
C-9.	All Flights over GCNP	6
E-1.	Relative Location of Measurement Sites in Grand Canyon	2
E-2.	Elevation Profile: Hancock Knoll to Swamp Point	3
E-3.	Hancock Knoll Measurement and Camp Sites	3
E-4.	Swamp Point Measurement Site	4
E-5.	Hancock Knoll Instrumentation, Part 1	4
E-6.	Hancock Knoll Instrumentation, Part 2	5
E-7.	Swamp Point Measurement Site	5
E-8.	Example Time Histories Measured at Hancock Knoll	7
E-9.	Representative A320 Aircraft Maximum Spectra Measured at Hancock Knoll	8
E-10.	Hancock Knoll Sound Level Histogram – All Data	9
E-11.	Hancock Knoll Sound Level Histogram – 7/16/2004, p.m.	9
E-12.	Hancock Knoll Sound Level Histogram – 7/17/2004, early p.m.	10
E-13.	Hancock Knoll Sound Level Histogram – 7/17/2004, late p.m.	10
E-14.	Hancock Knoll Sound Level Histogram – 7/18/2004, a.m.	11
E-15.	Hancock Knoll Sound Level Histogram – 7/18/2004, p.m.	11
E-16.	Hancock Knoll Sound Level Histogram – 7/19/2004, a.m.	12
E-17.	Swamp Point Sound Level Histogram – All Data	12
E-18.	Swamp Point Sound Level Histogram – 7/17/2004, early p.m.	13
E-19.	Swamp Point Sound Level Histogram – 7/17/2004, late p.m.	13
E-20.	Swamp Point Sound Level Histogram – 7/18/2004	14
E-21.	Swamp Point Sound Level Histogram – 7/19/2004	14
E-22.	Hancock Knoll Natural Sound Duration Histogram – All Data	15
E-23.	Hancock Knoll Natural Sound Duration Histogram – 7/16/2004	15
E-24.	Hancock Knoll Natural Sound Duration Histogram – 7/17/2004, early p.m.	15
E-25.	Hancock Knoll Natural Sound Duration Histogram – 7/17/2004, late p.m.	16
E-26.	Hancock Knoll Natural Sound Duration Histogram – 7/18/2004, a.m.	16
E-27.	Hancock Knoll Natural Sound Duration Histogram – 7/18/2004, p.m.	16
E-28.	Hancock Knoll Natural Sound Duration Histogram – 7/19/2004	17
E-29.	Swamp Point Natural Sound Duration Histogram – All Data	17
E-30.	Swamp Point Natural Sound Duration Histogram – 7/17/2004	17

Table of Figures (continued)

<u>Figure</u> <u>Page</u>

E-31. Swamp Point Natural Sound Duration Histogram – 7/18/2004 18
E-32. Swamp Point Natural Sound Duration Histogram – 7/19/2004 18
E-33. Summary of Concurrent Acoustic States 19

List of Acronyms

ADR	Alternative Dispute Resolution
AEE	Office of Environment and Energy
AIR	Aerospace Information Report
ANSI	American National Standards Institute
ARP	Aerospace Recommended Practice
ASQP	Airline Service Quality Performance
ASCII	American Standard Code for Information Interchange
ATMP	Air Tour Management Plan
CAEP	Committee on Aviation Environmental Protection
CFDR	Digital Flight Data Recorder data
COE	Center of Excellence
DNL	Day Night Average Sound Level
DEM	Digital Elevation Model terrain data
DLG	Digital Line Graph mapping data
DoD	Department of Defense
DoI	Department of Interior
DOT	Department of Transportation
DRG	Design Review Group
EASN	Equivalent Auditory System Noise
ECAC	European Civil Aviation Conference
EPNL	Effective Perceived Noise Level
ETMS	Enhanced Traffic Management System
FAA	Federal Aviation Administration
FAR	Federal Aviation Regulations
FICAN	Federal Interagency Committee on Aircraft Noise
GA	General Aviation
GB	Gigabyte
GCNP	Grand Canyon National Park
GNU GPL	GNU's Not UNIX General Public License
GTD	Geometric Theory of Diffraction
GUI	Graphical User Interface
HNK	Hancock Knoll (Grand Canyon)
HNM	Heliport Noise Model
ICAO	International Civil Aviation Organization
INM	Integrated Noise Model
ISO	International Organization for Standardization
JPDO	Joint Planning and Development Office
LaRC	Langley Research Center
LOS	Line of Sight
MIL	Military
MVS	Model Validation Study
NASA	National Aeronautics and Space Administration
NATO-CCMS	North Atlantic Treaty Organization Committee on the Challenges of Modern Society
NIRS	Noise Integrated Routing System

List of Acronyms (continued)

NMSim	Noise Model Simulation; also Noise Map Simulation
NOAA	National Oceanographic and Atmospheric Administration
NPD	Noise Power Distance
NPS	National Park Service
OAG	Official Airline Guide
PDARS	Performance Data Analysis and Reporting System (http://pdars.arc.nasa.gov/)
RNM	Rotorcraft Noise Model
SAE	Society of Automotive Engineers
SAGE	System for Assessing Aviation's Global Emissions
SEL	Sound Exposure Level
SOW	Statement of Work
SWP	Swamp Point (Grand Canyon)
TA	Time Above
TOR	Terms of Reference

1. Introduction

In a letter to Mr. Alan Zusman, the Chairperson of the Federal Interagency Committee on Aviation Noise (FICAN), dated September 2, 2003, the Federal Aviation Administration (FAA) and National Park Service (NPS) jointly requested that FICAN "provide advice on some matters related to the measurement and assessment of the effects of aircraft noise due to overflights of units of the National Park System". Accompanying the letter was a mutually agreed upon FAA/NPS Terms of References (ToR) document and a general Statement of Work (SOW). The SOW calls for the conduct of a comprehensive review of available computer models to be used for assessing aircraft noise in Grand Canyon National Park (GCNP), as well as in other National Parks. The letter, ToR and SOW are all included as background in Appendix A of this document.

At a September 17, 2003, meeting FICAN agreed to assist the FAA and NPS. FICAN then enlisted the assistance of the U.S. Department of Transportation's Volpe Center (Volpe) and Wyle Laboratories (Wyle) to assist with the study. Volpe is responsible for the development of the core acoustics module within the FAA's Integrated Noise Model (INM), and Wyle is responsible for the development of the Department of Defense's (DoD) NoiseMap SIMulation model (NMSim).

On October 29, 2004, FICAN met with members of Volpe and Wyle to discuss the results of the study to date. The discussions focused on the draft report dated October 21, 2004. At the conclusion of the October 29 meeting, FICAN concluded that there was not sufficient information to support a definitive finding. Although the two models were shown to perform equally well when compared with "gold standard" GCNP field measured data, there was a large enough difference when comparing the output of the two models to warrant further investigation. Consequently, FICAN requested that the Volpe/Wyle team focus additional studies on better understanding the differences between the two models.

The scope of the study is limited to the latest versions of INM (Version 6.2) and NMSim (Version 3.0). Volpe and Wyle worked cooperatively in the conduct of all analyses supporting this effort, including the layout and drafting of this report.

1.1 Study Background and Introduction to the Models

In January 2003, the NPS released Reference 1, which lays out in detail a comprehensive noise model validation study undertaken jointly in 1999 by the FAA and NPS at GCNP. Included in Reference 1 (among other things) is a detailed statistical assessment of the performance of a special research version of the INM (circa 1999) and NMSim (Version 2.3A, circa 1999). The document concluded that NMSim was the model of choice for conductance of air-tour noise analyses in GCNP, as well as in other parks. As part of the statistical analysis, the document cited specific areas of improvement for all models evaluated, including the research version of INM as well as for NMSim, e.g., it indicated that both models would benefit from the inclusion of an algorithm capable of accounting for propagation through dense vegetation, such as trees. It also cited that a potential area of improvement for INM would be the ability to account for shielding of the source-to-receiver propagation path by terrain, a particular issue in GCNP, as well as in other parks.

As a result of these recommendations, substantial enhancements were made to the INM core acoustics module. These enhancements specifically address many of the unique requirements associated with modeling in a National Park environment, including the ability to account for terrain shielding and an upgrade to the model to support higher fidelity terrain data. In addition, the INM's core noise and performance database was substantially expanded to include many of the tour aircraft common in a National Park environment. A detailed summary of the enhancements included in INM to specifically address the needs of the National Parks' modeler are presented in Appendix B.

During the same period, NPS commissioned the development of NMSim from an engineering-oriented DOS program into a user-friendly GUI Windows program. The updated program, denoted "Noise Model Simulation" is the current version of NMSim, and is planned for release at the conclusion of the FICAN study. In addition to the user-friendly interface, it contains improvements in database, geocoding and other infrastructure. Core noise calculations are unchanged from Version 2.3A.

1.1.1 INM

The FAA's INM, originally released in 1978, is the most widely distributed aircraft noise prediction tool in the world – it has over 800 users in more than 40 countries. The FAA's Office of Environment and Energy (AEE) developed the model, with the assistance of the ATAC Corporation, which acts as systems integrator, and Volpe is responsible for the development and enhancement of the core acoustics. The National Aeronautics and Space Administration (NASA) has also contributed substantially to the advancement of the core acoustics and the database within the model. INM has been continually updated, with over six major releases since its inception, along with dozens of minor releases (see Figure 1). An international design review group (DRG) has also largely influenced the development of the model. The INM DRG is made up of a body of users from government, industry and academia. In addition, the model currently adheres to numerous international technical standards [2, 3, 4, 5], further supporting its viability in a public process; the model is currently being upgraded for adherence to the newly developed aircraft noise modeling standard of the European Union [6].

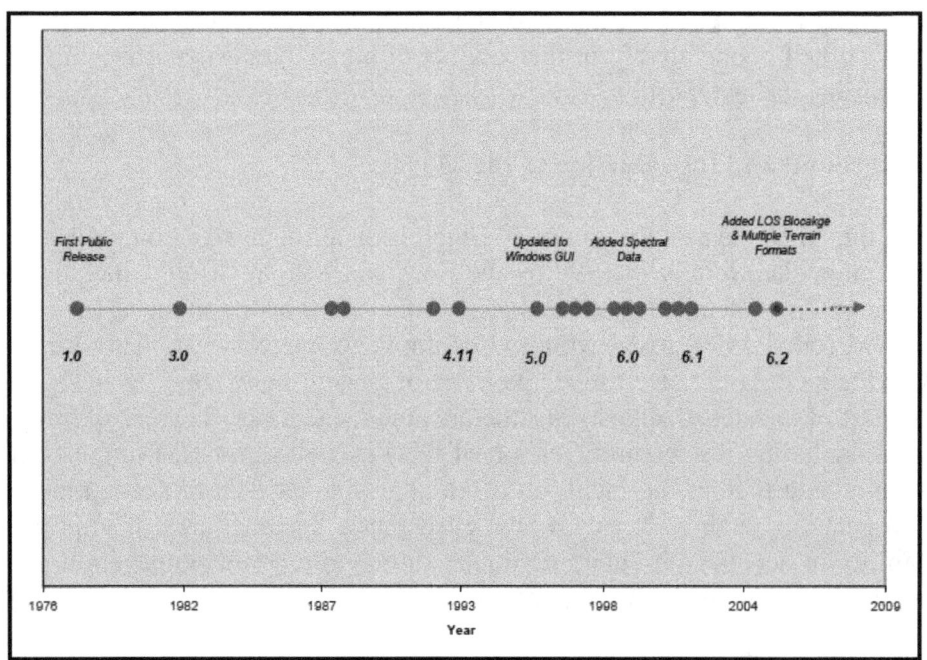

Figure 1. INM Development Timeline

With regard to basic physics, INM is considered a line source model, with one-third octave-band-based core acoustic computations. The fundamental computations take into account divergence, atmospheric absorption, terrain shielding and ground effects. In addition to noise computations, the model also includes a detailed aircraft performance module, which is essential to precise aircraft noise prediction [7]. The model includes the ability to account for performance in the terminal area, as well as enroute performance at low to moderate altitudes, as is the case for air tours in the National Parks. The INM also maintains a comprehensive noise-power-distance and associated aircraft performance database, which is continually augmented with input from aircraft manufacturers, as well as through supplementary FAA- and NASA-sponsored field measurement studies [8-27]. INM computations are facilitated by a user-friendly, Windows-based graphical user interface. A dbf file structure also allows easy, external manipulation of the model's input/output data [28]. As a publicly available tool, FAA offers the INM user community free and timely technical support. In addition, several private firms offer periodic INM training.

1.1.2 NMSim

NMSim (Noise Model Simulation) [29] is a noise simulation model [30] that evolved from a NATO-CCMS study on the effects of topography on sound propagation around airfields [31]. Its gestation was analysis of noise from an international propagation experiment [32], and evaluation of ray tracing sound propagation models [33,34,35,36]. It evolved into a full one-third octave simulation model based on three-dimensional sources [37], with its initial application [38] being R&D support for DoD's NoiseMap [39] airbase noise model. Successful validation of propagation algorithms via the Narvik experiment [32, 41] provided support for implementation of topography algorithms in NoiseMap 7. It was subsequently developed into a self-standing model used by DoD [40] and NASA [41,42]. It has also been used by Wyle in projects for various clients. A useful feature of NMSim is that, as a full simulation model, it is capable of generating color animations of noise from moving sources.

NMSim was built for analysis of propagation over terrain. It has a modular structure, and special versions have been employed to assess the effects of meteorology on airport/airbase noise [43,44]. NMSim is closely related to the Wyle/NASA developed RNM (Rotorcraft Noise Model) that is used by NASA, DoD, the helicopter industry and NATO partners for analysis of rotorcraft noise. The primary difference between NMSim and RNM is that RNM incorporates complex multi-component noise sources (e.g., tiltrotors) while NMSim assumes compact sources as traditionally formulated for fixed-wing aircraft and simplified representation of rotorcraft.

NMSim Version 2.3A, as used in the GCNP MVS, was DOS-based, and had limited tools for setting up cases. Following its application in the MVS, NPS sponsored development into the current user-friendly Graphical User Interface 32 bit Windows application, NMSim Version 3.0 [29]. An international beta-test team comprised of noise experts from industry, consultancies, military, and government agencies (including FAA, DOT and NASA), supported that development. The NPS version of NMSim is distributed under the GNU General Public License. Should the need for training courses arise, Wyle plans on offering such services.

1.2 Objectives

The first objective of this study was to evaluate the series of model enhancements that were included in INM as a result of the recommendations from the GCNP MVS. Specifically, there was a desire to evaluate the performance of the latest versions of INM and NMSim[1], compared with the "gold standard" data measured in the GCNP MVS. The second, but equally important objective of the current study was to examine the issue of model usability, e.g., ease of operation, runtime, data input/availability, etc. The issue of usability is of particular importance within the context of Air Tour Management Plans (ATMPs), as the development of ATMPs is a public process, which will require noise modeling in well over 100 National Parks. Additionally, as a result of a court ruling regarding environmental studies associated with St. George Airport in Utah, the courts identified the requirement to also consider the cumulative effects of noise from all aviation sectors on a National Park, including the effects of high altitude jet aircraft. Hence, the third objective of the current study was to assess the applicability of the two models with regard to assessing noise from high altitude jet aircraft. In support of the third objective a field measurement study was also conducted. The details of the measurement study are presented in Appendixes C, D, and E.

1.3 Organization of Document

Section 1 of this report presents an introduction to the study, including a basic description of the two models being evaluated, the FAA's Integrated Noise Model (INM) and DoD's NoiseMap Simulation Model (NMSim). Section 2 presents a systematic comparison of the two models, including a comparison of the physics, and the underlying databases. Section 3 presents a series of parametric studies/comparisons of the two models, along with an updated statistical assessment of their latest versions compared with the measured data collected in GCNP in 1998. Section 4 presents a comparative assessment of the two models within the context of an actual case study in GCNP. Also included in this section is a detailed assessment of model sensitivities. Section 5 addresses the topic of model usability. Section 6 presents a summary of the study findings and recommended improvements to the models. The report also includes several supporting appendices that provide additional detail.

---

[1] When this study was originally undertaken, it was intended that the core acoustics in NMSim would not be modified from Version 2.3A, as used in the MVS. As a result of the latest round of evaluations (post October 2004 meeting), the audibility module in both INM and MNSim were updated to ensure consistency.

2. Comparison of Model Capabilities

Table 1 provides a summary comparison of the main computational components within INM and NMSim. This section is laid out in a manner similar to that presented in the table. The table compares the underlying noise database within the two models (noted as "database" in the table; Section 2.1). It also addresses the fundamental physics of INM and NMSim ("physics"; Section 2.2). Additionally, the table compares other computational capabilities within the two models – such as noise contouring ("other").

Table 1. Comparison of Features in INM and NMSim

Capability	INM	NMSim
Minor Differences Between Models		
Terrain Data - *other*	3CD, DEM, GridFloat (may be user-defined)	DLG, DEM, DTED; open format
Atmospheric Absorption - *physics*	SAE-ARP-866A	ANSI/ISO 9613
Case History	800+ Users World-wide over 30 Years	In use since 1996 by DoD, NASA, NATO, Wyle, manufacturers
Mixed Ground Impedance - *physics*	Hard or Soft; Research Version: Fresnel Zone-based distance weighting	Hard/Soft proportional distance weighting
Ground/Terrain Effects - *physics*	FHWA (Maekawa - Kurze/Anderson)	Rasmussen full topography
Highway Noise Sources - *other*	Merge FHWA TNM model output with NM output	Aggregate noise hemisphere based on FHWA TNM
Hard Ground - *physics*	Select either "Hard" or "Soft" (SAE-AIR-1751); Research Version: Import hydrography	Import hydrography for water sources; augment for other "hard" surfaces
Source Code - *other*	Available for Researchers only	*will be available* under GNU Gerneral Public License
Source Code Language - *other*	C++	Fortran
Aircraft Bank Angle - *other*	Version 7 0, ECAC Doc 29R	Yes
One-Third Octave Band Coverage - *database*	Standardized 50 Hz to 10 kHz for all data	25 Hz to 10 kHz for most data. Frequency range user defined as needed
Database User Accessibility - *database*	All standardized; ability to add user-defined aircraft and profiles	All data accessible to users
Noise Descriptors - *other*	Standard (A-, C- and tone-corr): SEL, DNL, CNEL, LEQ, LAeq(Day), LAeq(Night), Lmax, (%)TA, Ddose; (%)TAUD. + user-defined versions of all metrics.	Flat-, A- and C- Max and Exposure. Leq(24), Ldn, TAUD, TA, DNL. Full spectral time histories
Change in Exposure Noise Metric - *other*	Yes	External to model
Models Consistent		
Interpolation/Extrapolation - *physics*		Consistent with NOISEMAP
One-Third Octave Band Effects - *physics*		Evaluated at center frequencies
Use of Multi-Resolution Terrain Data - *other*		Okay
Major Differences Between Models		
Noise Database Structure - *database*	NPD data as a function of power (P)	Noise at each power and operational configuration defined on sphere
Database Coverage - *database*	115 commercial; 110 military (from NOISEMAP); 28 turboprop/piston; 17 helicopters	6 GCNP aircraft (in-situ, low elevation angles, as measured by FAA); 3 military; 39 INM-derived aircraft; 1 generic rail source; generic highway sources from TNM 2.0. Can import any aircraft from Noisefile or NM database
Database Development - *database*	Manufacturers continually adding/updating per SAE-AIR-1845	Military aircraft added as needed. dB towers *will* permit production gathering of noise source data (to be half constructed in CY2005; hope for further construction in CY2006)
Overlapping Time Histories - *other*	Research Version or external to model	Yes
Aircraft Performance - *other*	SAE-AIR-1845	User input; can import NM fixed-point profiles

2.1 Source Characterization

An important component of the two models is the underlying noise level database. The noise databases in INM and NMSim are both founded in field measurements, but the overall structure of the data in the two models is different. Section 2.1.1 discusses in detail the overall differences in the structure of the two databases. Section 2.1.2 compares the coverage of the database in each model.

2.1.1 Description

The noise level database in INM is often referred to as the noise-power-distance (NPD) database. For each aircraft represented, the INM contains noise level data as a function of distance (200 to 25,000 ft) for a range of representative aircraft power settings – from approach through full power takeoff. Figure 2 graphically presents the NPD data set included in INM for the DeHaviland DHC-6 aircraft configured with the Raisbeck quiet propellers, a common turbo-prop aircraft used for air tours in many of the larger U.S. National Parks.

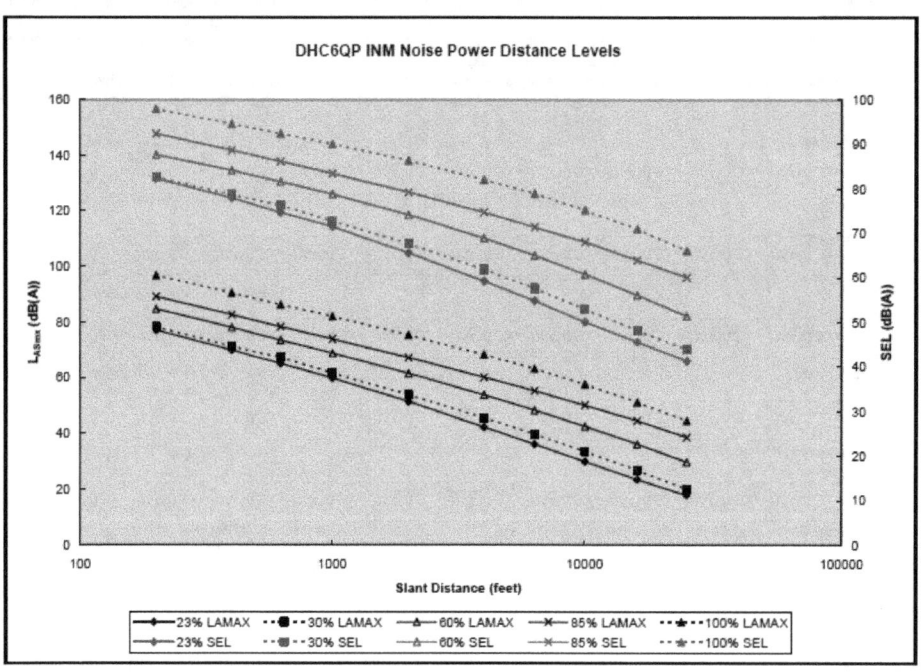

Figure 2. Example NPD Data for DeHaviland DHC-6 with Quiet Propellers

For each aircraft type, the INM database contains NPD data sets for up to four basic noise metrics, each representing the four fundamental metrics from which all other metrics in the model are computed. These noise metrics are the sound exposure level (SEL, denoted by the symbol L_{AE}), the maximum A-weighted sound level (MXSA, denoted by the symbol L_{ASmx}), the effective perceived noise level (EPNL, denoted by the symbol L_{EPN}) and the tone-corrected, maximum perceived sound level (MXSPNT), denoted by the symbol L_{PNTSmx}). For fixed-wing aircraft, the NPD data set for each aircraft is representative of a flight passing directly overhead. Lateral source directivity effects for fixed-wing aircraft are not accounted for in the INM database, but rather in the lateral effects algorithm (see Section 2.2.2). For helicopters, each flight configuration is represented by three data sets, a center, a left and a right data set – representing the noise signature directly below, to the left- and to the right-side of the helicopter. The INM interpolates between these three data sets to take into account lateral directivity effects from helicopters. Forward and aft directivity in INM (for all aircraft types including helicopters) is based on a fourth-power dipole equation, augmented by a special directivity function for aircraft behind the start-of-takeoff roll.

One-third octave band data for INM are based upon 'spectral classes'. The spectral classes are based on the operating state of the aircraft and are defined as 'Arrival', 'Departure', 'Fly-Over' and/or 'Afterburner' (for military jets). The actual spectra (as opposed to spectral classes) for the various operating conditions may be considered proprietary by the aircraft manufacturers and so are unavailable for dissemination with the model.

The INM benefits from the fact that the primary commercial aviation manufacturers (e.g., Airbus, Boeing, Embraer, etc.) provide noise level (and performance – see Section 2.5 below) data directly to the FAA for inclusion in the INM. In fact, efforts are ongoing through a joint collaboration between FAA and Eurocontrol to adopt the INM database as the internationally accepted aircraft noise level data base to be used for aviation noise modeling. The significance of this initiative is that it will result in the INM database being available online, continually updated and maintained, and being generally accepted as best practice worldwide.

In addition to the manufacturers' data, the FAA and NASA have funded a series of field measurement efforts to augment the INM database, particularly for the smaller aircraft and helicopters that are more common to a National Park environment. [8-27]

The structure of the noise level database in NMSim is similar to that in INM: noise is defined at a single speed and several power settings, with noise at other power settings interpolated. There are two differences:

- In NMSim, the noise from the aircraft is defined in terms of one-third octave band sound levels as a function of emission angles, i.e., in terms of a sphere of noise, rather than an integrated SEL value.

- Source noise is defined at one distance. Noise at other distances is computed within the program. If sound exposure level (SEL) is computed by NMSim, the process is equivalent to the SAE AIR 1845 Type 1 procedure used to develop INM's A-weighted NPD curves from its original spectral data.

NMSim's noise database is not as comprehensive as INM's. Ideally, noise spheres are prepared from flight test data by the Wyle/NASA ART2 process [46]. That has been accomplished for several military aircraft and rotorcraft. Partial noise spheres have been prepared for the six aircraft modeled in the MVS [1], using data collected during that study. A procedure has been developed to prepare NMSim noise sources from the INM A-weighted NPD database and spectral classes. Sources derived from INM NPDs are considered to be low resolution relative to the ART2 process because the available NPD data have been integrated into SEL and only spectral class data are available. Most of the original spectral time history data forming the INM database are proprietary to the aircraft manufacturers, and are not available to NMSim.

NMSim does not model aircraft performance. It contains GUI tools for entering flight paths, but it is up to the user to define performance. Following NoiseMap heritage, NMSim flight paths combine track and profile data. Flight paths can be generated by importing INM tracks and profiles.

2.1.2 Fleet Coverage

The INM NPD database is by far the most comprehensive set of aircraft noise data in the world. It includes data for 253 fixed-wing total aircraft, including 115 commercial aircraft, 110 military aircraft, 28 small turboprop and piston aircraft and 17 helicopters. Figure 3 presents the coverage of the INM Version 6.1 NPD database in terms of the percentage of the active aviation fleet (based on the aircraft in the Official Airlines Guide, OAG). As can be seen, the INM NPD database represents about forty percent of the active fleet. That number is continually increasing; currently there are approximately twenty INM noise and performance database projects underway. Table 2 presents the aircraft represented in the INM NPD database. For aircraft not represented directly in the INM, the model also contains a comprehensive list of appropriate aircraft substitutions. The INM database submittal form and process ensure a tight linkage that allow for stringency and operational procedure analysis.

In support of this study, a comprehensive review of the ATMP Interim Operating Authority documents provided by the operators as of March 2004 indicated that data for five additional aircraft types would provide complete coverage in the INM database with regard to modeling in the National Parks: Beech C99, Cessna 182, Cessna 208, Fokker F-27, and the Robinson R44. The Beech C99 is currently modeled in the INM with the substitution aircraft type of the DHC6. The C99 is a significantly faster aircraft, this speed difference impacts the accuracy of the Time Above and Time Audible metrics. C99s operate over 13 Parks. The Cessna 182 is currently modeled with the CNA206 (Cessna 206) substitution; earlier versions of the C182 use an engine with faster propeller rotational speed than later versions. Flight tests have shown that propeller-driven aircraft noise is a function of propeller rotational speed. Cessna 182s operate over 50 Parks. The Cessna 208 is currently modeled with the GASEPF (General Aviation Single Engine Pitch Fixed) substitution; the GASEPF is intended to model the lightest and quietest reciprocating-engine aircraft in the fleet; the C208 is a relative large propeller aircraft with a turbine engine. Given the different spectral signatures of the two aircraft types, the C208 may not be well modeled. The C208 operates over 3 Parks. The Fokker F-27 was withdrawn from service last year due to inability to met current security requirements. There is no longer a need to add this aircraft to the database. The Robinson R-44 helicopter is currently modeled in the INM with the Hughes 500 (H500D) substitution. The Hughes 500 is a five bladed turbine-powered helicopter, while the R44 is a two-bladed reciprocating-engine helicopter. The noise signatures of these two aircraft can be expected to be significantly different. The R44 operates over seven Parks.

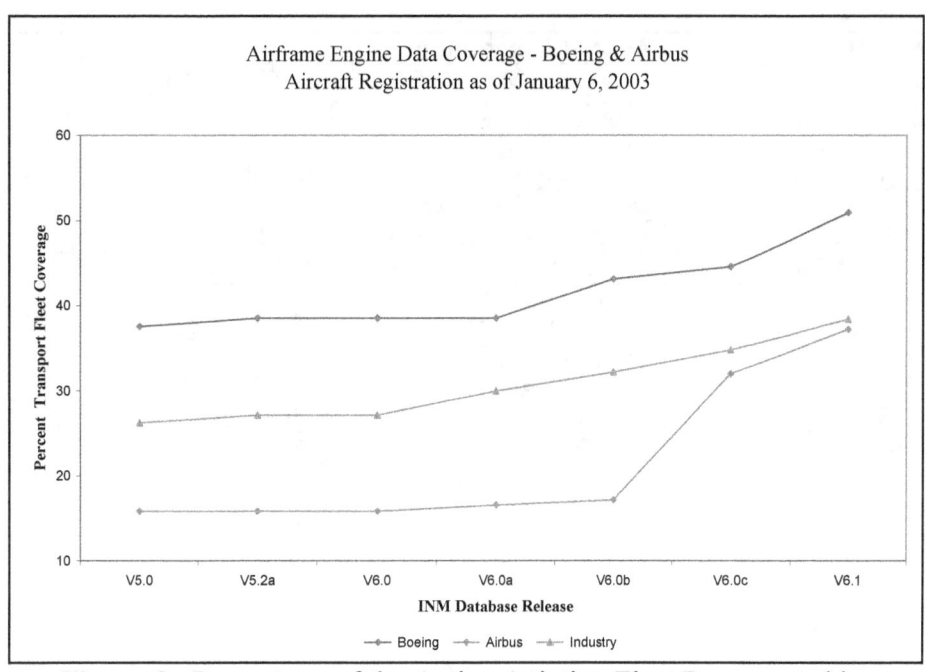

Figure 3. Percentage of the Active Aviation Fleet Represented by the INM NPD Database

Table 2. Aircraft Represented in the INM NPD Database

INM Aircraft ID	Aircraft Description
Commercial	
1900D	Beech 1900D / PT6A67
737800 / 757300	Boeing 737-800/CFM56-7B26 // 757-300/RB211-535E4B
717200 / 777300	Boeing 717-200/BR 715 // 777-300/TRENT892
707 / 707120 / 707320	707-120/JT3C // 707-120B/JT3D-3 // 707-320B/JT3D-7
707QN / 720	Boeing 707-320B/JT3D-7QN // 720/JT3C
720B / 727100 / 727200	Boeing 720B/JT3D-3 // 727-100/JT8D-7 // 727-200/JT8D-7
727D15 / 727D17 / 727EM1	727-200/JT8D-15 // 727-200/JT8D-17 // FEDX 727-100/JT8D-7
727EM2 / 727Q15 / 727Q7	727-200/JT8D-15 // 727-200/JT8D-15QN // 727-100/JT8D-7QN
727Q9 / 727QF	Boeing 727-200/JT8D-9 // UPS 727100 22C 25C
737 / 737300 / 7373B2	Boeing 737/JT8D-9 // 737-300/CFM56-3B-1 // 737-300/CFM56-3B-2
737400 / 737500 / 737D17	B 737-400/CFM56-3C-1 // 737-500/CFM56-3B-1 // 737-200/JT8D-17
737QN	Boeing 737/JT8D-9QN
747100 / 74710Q / 747200	B 747-100/JT9DBD // 747-100/JT9D-7QN // 747-200/JT9D-7
74720A / 74720B / 747400	B 747-200/JT9D-7A // 747-200/JT9D-7Q // 747-400/PW4056
747SP	Boeing 747SP/JT9D-7
757PW / 757RR	Boeing 757-200/PW2037 // 757-200/RB211-535E4
767300 / 767CF6 / 767JT9	B 767-300/PW4060 // 767-200/CF6-80A // 767-200/JT9D-7R4D
767CF6 / 767JT9	Boeing 767-200/CF6-80A // 767-200/JT9D-7R4D
A300	Airbus A300B4-200/CF6-50C2
BAC111 / BAE146	BAC111/SPEY MK511-14 // BAE146-200/ALF502R-5
BAE300	BAE146-300/ALF502R-5
CNA441 / CONCRD	CONQUEST II/TPE331-8 // CONCORDE/OLY593
CVR580	CV580/ALL 501-D15
DC1010 / DC1030 / DC1040	DC10-10/CF6-6D // DC10-30/CF6-50C2 // DC10-40/JT9D-20
DC3 / DC6	DC3/R1820-86 // DC6/R2800-CB17
DC820 / DC850 / DC860	DC-8-20/JT4A // DC8-50/JT3D-3B // DC8-60/JT3D-7

INM Aircraft ID	Aircraft Description
DC870 / DC8QN	DC8-70/CFM56-2C-5 // DC8-60/JT8D-7QN
DC910 / DC930 / DC950	DC9-10/JT8D-7 // DC9-30/JT8D-9 // DC9-50/JT8D-17
DC9Q7 / DC9Q9	DC9-10/JT8D-7QN // DC9-30/JT8D-9QN
DHC6 / DHC7 / DHC8	DASH 6/PT6A-27 // DASH 7/PT6A-50 // DASH 8-100/PW121
DHC830	DASH 8-300/PW123
F10062 / F10065	F100/TAY 620-15 // F100/TAY 650-15
F28MK2 / F28MK4	F28-2000/RB183MK555 // F28-4000/RB183MK555
HS748A	HS748/DART MK532-2
L1011 / L10115	L1011/RB211-22B // L1011-500/RB211-224B
L188	L188C/ALL 501-D13
MD11GE / MD11PW	MD-11/CF6-80C2D1F // MD-11/PW 4460
MD81 / MD82 / MD83	MD-81/JT8D-209 // MD-82/JT8D-217A // MD-83/JT8D-219
SD330 / SF340	SD330/PT6A-45AR // SF340B/CT7-9B
MD9025 / MD9028	MD-90/V2525-D5 // MD-90/V2528-D5
737N17 / 737N9	B737-200/JT8D-17 Nordam B737 LGW Hushkit // B737/JT8D-9
777200	Boeing 777-200ER/GE90-90B
DC93LW / DC95HW	DC9-30/JT8D-9 w/ ABS Lightweight hushkit // DC9-50/JT8D17
EMB145 / EMB14L	Embraer 145 ER/Allison AE3007 // 145 LR / Allison AE3007A1
DHC6QP	DASH 6/PT6A-27 Raisbeck Quiet Prop Mod
A340	Airbus A340-211/CFM 56-5C2
EMB120	Embraer 120 ER/ Pratt & Whitney PW118
A320 / A330	Airbus A320-211/CFM56-5A1 // A330-301/CF6-80 E1A2
737700 / 767400	Boeing 737-700/CFM56-7B24 // 767-400ER/CF6-80C2B(F)
A319 / A32023	Airbus A319-131/V2522-A5 // A320-232/V2527-A5
A33034 / A32123	Airbus A330-343/RR TRENT 772B // A321-232/IAE V2530-A5
A310 / A30062	Airbus A310-304/CF6-80C2A2 // A300-622R/PW4158
CNA750	Citation X / Rolls Royce Allison AE3007C
BEC58P	BARON 58P/TS10-520-L
CIT3 / CL600 / CL601	CIT 3/TFE731-3-100S // CL600/ALF502L // CL601/CF34-3A
CNA500	CIT 2/JT15D-4
COMJET / COMSEP	1985 BUSINESS JET // 1985 1-ENG COMP
FAL20	FALCON 20/CF700-2D-2
General Aviation	
GASEPF / GASEPV	1985 1-ENG FP PROP // 1985 1-ENG VP PROP
IA1125	ASTRA 1125/TFE731-3A
LEAR25 / LEAR35	LEAR 25/CJ610-8 // LEAR 36/TFE731-2
M7235C / MU3001	MAULE M-7-235C / IO540W // MU300-10/JT15D-4
SABR80	NA SABRELINER 80
CNA172 / CNA206	Cessna 172R / Lycoming IO-360-L2A // 206H / IO-540-AC
CNA20T	Cessna T206H / Lycoming TIO-540-AJ1A
CNA55B	Cessna 550 Citation Bravo / PW530A
GII / GIIB	Gulfstream GII/SPEY 511-8 // Gulfstream GIIB/GIII- SPEY 511-8
GIV / GV	Gulfstream GIV-SP/TAY 611-8 // Gulfstream GV/BR 710
PA28	PIPER WARRIOR PA-28-161 / O-320-D3G
PA30	PIPER TWIN COMANCHE PA-30 / IO-320-B1A
PA31	PIPER NAVAJO CHIEFTAIN PA-31-350 / TIO-5
A7D	A-7D,E/TF-41-A-1
C130 / C130E	C-130H/T56-A-15 // C-130E/T56-A-7
KC135 / KC135B / KC135R	KC135A/J57-P-59W // KC135B/JT3D-7 // KC135R/CFM56-2B-1
F4C	F-4C/J79-GE-15
Military	
A10A	FAIRCHILD THUNDERBOLT II TF34-GE-100 NM
A37	CESSNA DRAGONFLY J85-GE-17A NM

INM Aircraft ID	Aircraft Description
A3 / A4C	MCDONNELLDOUGLASSKYWARRIORJ79-GE-8NM//J52-P-8A
A5C / A6A	J79-GE-10 NM // GRUMMAN INTRUDER J52-P-8B NM
A7E	VOUGHT CORSAIR II TF41-A-2 NM
AV8A / AV8B	BAE HARRIER AV8A NM // BAE HARRIER F402-RR-408 NM
B1	ROCKWELL LANCER F101-GE-102 NM // F118-GE-110 NM
B52BDE	STRATOFORTRESS J57P-19W NM // J57-P-43WB NM
B52H	BOEING STRATOFORTRESS B52H NM
B57E	ENGLISH ELECTRIC CANBERRA J57-PW-P-5 NM
BUCCAN	RR SPEY RB 168-1A NM
C-130E	LOCKHEED HERCULES T56-A15 C130E NM
C-20	GULFSTREAM III MK611-8RR NM
C118	MCDONNELL DOUGLAS LIFT PW R-2800-CB17 NM
C119L	FAIRCHILD FLYING BOX CAR C119L NM
C121 / C123K	C121 NM // FAIRCHILD PW R-2800-99W AUX J85-GE17 NM
C12	BEECH SUPER KING AIR HURON PW PT6A-41 NM
C130AD / C130HP	LOCKHEED HERCULES T56-A15 NM // C130HP NM
C131B	GENERAL DYNAMICS CV34 PW R-2800-99W NM
C135A / C135B	BOEING STRATOLIFTER PW J57-59W NM // C135B NM
C137	JT3D-3B NM
C140 / C141A	LOCKHEED JETSTAR TFE731-3 NM// STARLIFTER TF-33-P-7
C17 / C18A	F117-PW-100 NM // JT41-11 NM
C21A	LEARJET 35 TFE731-2-2B NM
C22 / C23	BOEING 727 TRS18-1 NM // PT6A-65AR NM
C5A	LOCKHEED GALAXY TF39-GE-1 NM
C7A	DEHAVILLAND CARIBOU DHC-4A NM
C9A	MCDONNELL DOUGLAS DC9 JT8D-9 NM
CANBER	2 RR AVON 109 NM
DOMIN	BRISTOL SIDDELEY VIPER 521 NM
E3A / E4	BOEING SENTRY TF33-PW-100A NM // B 747 CF6-50E NM
E8A / EA6B	JT3D-3B NM // J52-P-408 NM
F-18	MCDONNELL DOUGLAS HORNET F404-GE-400 NM
F-4C	MCDONNELL DOUGLAS PHANTOM J79-6517A17 NM
F100D	ROCKWELL SUPER SABRE PW J57-P-21A NM
F101B / F102	PW J57-P-55 NM // PW J57-P-23 NM
F104G	LOCKHEED STARFIGHTER J79-GE-11A NM
F105D / F106	PW J75-P-19W NM // PW J57-P-17 NM
F111AE / F111D	GENERAL DYNAMICS F111AE PW TF30-P-100 NM//F111D
F-111F / F117A	GENERAL DYNAMICS F111F NM // F404-GE-F1D2 NM
F14A / F14B	GRUMMAN TOMCAT TF30-P-414A NM // F110-GE-400 NM
F15A	MCDONNELL DOUGLAS EAGLE F100-PW-100 NM
F15E20 / F15E29	MCDONNELL F100-PW-220 NM // F100-PW-229 NM
F16A / F16GE	GENERAL DYNAMICS FALCON PW200 NM /F110-GE-100 NM
F16PW0 / F16PW9	GENERAL DYNAMICS F F100-PW-220 / F100-PW-229 NM
F18EF	Boeing F-18E/F / F414-GE-400 NM
F5AB / F5E	NORTHRUP TIGER J85-GE-13 NM // J85-GE-21B NM
F8	VOUGHT F-8 CRUSADER PW J57-P-201 NM
FB111A	GENERAL DYNAMICS FB111 PW TF30-P-100 NM
HARRIE	BAE HARRIER AV8 RR PEGASUS 6 NM
HAWK	RR ADOUR MK151 NM
HS748	RR DART RDA7 MK 536-2 NM
HUNTER / JAGUAR	RR AVON RA28 NM // SEPECAT JAGUAR NM
JPATS	Raytheon T-6A Texan II / PT6A-68 NM
KC-135	BOEING STRATOTANKER KC135R F108-CF100 NM

INM Aircraft ID	Aircraft Description
KC10A	CFG-50C2 NM
KC97L	BOEING STRATOFREIGHTER PW R-436-59B NM
LIGHTN / NIMROD	RR AVON 302C NM // RR SPEY MK511 NM
OV10A	ROCKWELL BRONCO T76 NM
P3A / P3C	LOCKHEED ORION T56-A-14 // T56-A-14 NM
PHANTO	MCDONNELL DOUGLAS PHANTOM F-4 NM
PROVOS	BRISTON SIDDELEY VIPER 11 NM
S3A&B / SR71	LOCKHEED VIKING TF34-6E-2 NM // JT11D-20B NM
T-2C	ROCKWELL BUCKEYE J85-6E-4 NM
T-38A / T-43A	NORTHRUP TALON T-38A NM // BOEING 737 T43A NM
T1	LOCKHEED SEA STAR JT15D-5 NM
T29	GENERAL DYNAMICS CV34 PW R-2800-99W NM
T33A	LOCKHEED T-33A J33-35 NM
T34	BEECH MENTOR (BE45) PT6A-25 NM
T37B	CESSNA 318 J69-T-25 NM
T39A / T3	ROCKWELL SABRELINER GEJ85 NM // AEIO-540-D4A5 NM
T41 // T42	CESSNA 172 O-320-E2D NM // BEECH BARON (BE55) NM
T44 // T45	T44 NM // PT6A-45AG NM
TORNAD / TR1	RB199-34R NM // J75-P-13B NM
U21 / U2	BEECH UTE PW PT6A-20 NM // LOCKHEED U2 J75-P-13 NM
U4B	ROCKWELL SUPER COMMANDER 1G0-540B1A NM
U6	DEHAVILLAND BEAVER PW R-985 DHC-2 NM
U8F	BEECH SEMINOLE 0-480-1 D50 NM
VC10	RR CONWAY RC0-42 NM
VICTOR	BRITISH AEROSPACE VICTOR NM
VULCAN	BRITTEN NORMAN VULCAN RR OLYMPUS 301 NM
YC14 / YC15	GE CF6-50D NM // PWJT8D-17 NM

The noise level database in NMSim currently includes:

- F-16, C-130 and Tornado processed via ART2 from original data recordings

- AS350, B206B, B206L, DHC6, C182, C207 processed via ART2 from GCNP MVS

- 7373B2, 737700, 747100, 747400, 77200, A320, 727200, 737300, 73717, 747200, 757300, BBARON, BEC58P, CNA206, CNA20T, CNA441, CNA441, CNA500, DHC6INM, DHC6QP, EC130, EMB145, GASEPF, GASEPV, MD83, PA31, SA350D, CIT3, CL600, CL601, GIIB, GV, GIV, HARRIE, IA1125, KC135R, MU3001, S3A&B, VC10 imported from INM

2.2 Propagation

The fundamental physics in the two models is generally consistent, but the specific formulations implemented in each are different. Section 2.2 compares and contrasts the physical propagation algorithms included in the two models.

2.2.1 Atmospheric Absorption

Atmospheric absorption in the two models is based on internationally accepted acoustic standards. The INM is based on the equations of SAE ARP 866A, while NMSim is based on those of ISO 9613, which is currently being reviewed by SAE for possible adoption in place of 866A. The equations in the two standards are quite similar, and in fact almost identical when computing absorption in the lower frequencies, which is the primary area of interest with regard to the National Parks due to the long propagation distances and the emphasis on time audible as a potentially preferred noise metric. Table 3 shows a comparison of the absorption coefficients by one-third octave-band for a temperature of 59 degrees (F) and relative humidity of 70 percent, as computed using the two methods.

Table 3. Comparison of Absorption Coefficients, SAE versus ISO[2]

One-Third Octave Band Nominal Center Frequency (Hz)	Atmospheric Absorption Coefficients (dB / 1,000 feet)	
	ISO	SAE
50	0.07	0.07
63	0.11	0.09
80	0.16	0.11
100	0.25	0.14
125	0.38	0.18
160	0.57	0.23
200	0.82	0.29
250	1.13	0.36
315	1.51	0.45
400	1.92	0.58
500	2.36	0.73
630	2.84	0.92
800	3.38	1.17
1000	4.08	1.47
1250	5.05	1.85
1600	6.51	2.39
2000	8.75	3.05
2500	12.20	4.02
3150	17.70	5.47
4000	26.40	7.64
5000	39.90	9.09
6300	61.10	12.80
8000	93.70	18.59
10000	144.00	27.44

Figure 4 presents the resultant NPD differences that result from atmospheric corrections using ISO and 866A. Data were derived using the spectra at time of A-weighted maximum sound level for the GCNP MVS source site data. As can be seen, there are some differences at the larger distances, but they are

---

[2] Atmospheric absorption coefficients for 59° F and 70 % RH.

generally less than 1 dB. It is worth noting that the largest difference occurs for the DHC-6 aircraft. This is discussed further in Section 3.1.

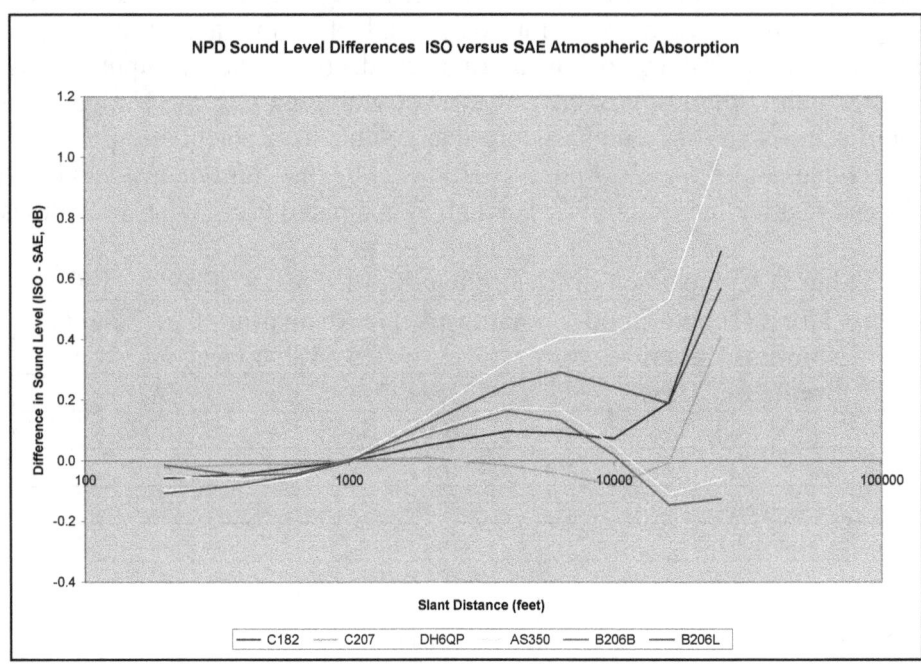

Figure 4. Summary of Sound Level Differences due to Differing Atmospheric Absorption

2.2.2 Lateral Effects

The lateral effects on aircraft propagation in INM are based on the soon-to-be-released update to SAE AIR 1751, *Lateral Effects on the Propagation of Aircraft Sound Levels*. The update is based on a series of international research studies conducted over the past six years [47-52]. It considers three primary acoustic phenomena: ground effects, meteorological effects (refraction and scattering) and lateral directivity.

The ground effects and refraction/scattering effects are independent of aircraft type in INM, i.e., sensitivity tests in support of standard development indicated that these effects were relatively independent of aircraft type and associated frequency spectra. However, the ground effects in INM are based on the ground surface characteristics: acoustically hard, acoustically soft, or a mix of both.

Often, a phenomenon such as lateral directivity would be considered a source characteristic and accounted for as part of the source characterization (see Section 2.1). This is the case in NMSim, but for many reasons, primarily historical, this effect in INM is folded into the algorithm used for computing lateral effects – at least for fixed-wing aircraft, as discussed in Section 2.1. In INM, two basic functions are used for computing lateral directivity. These functions, shown in Figure 5, are for jet aircraft with wing-mount (a) and tail-mount (b) engines. For propeller-driven aircraft, which are most relevant in a National Park environment, the lateral directivity function is set to zero. As mentioned above in Section 2.1, the lateral directivity function in INM for helicopters is implemented within the NPD database, because of the directional complexity of helicopters.

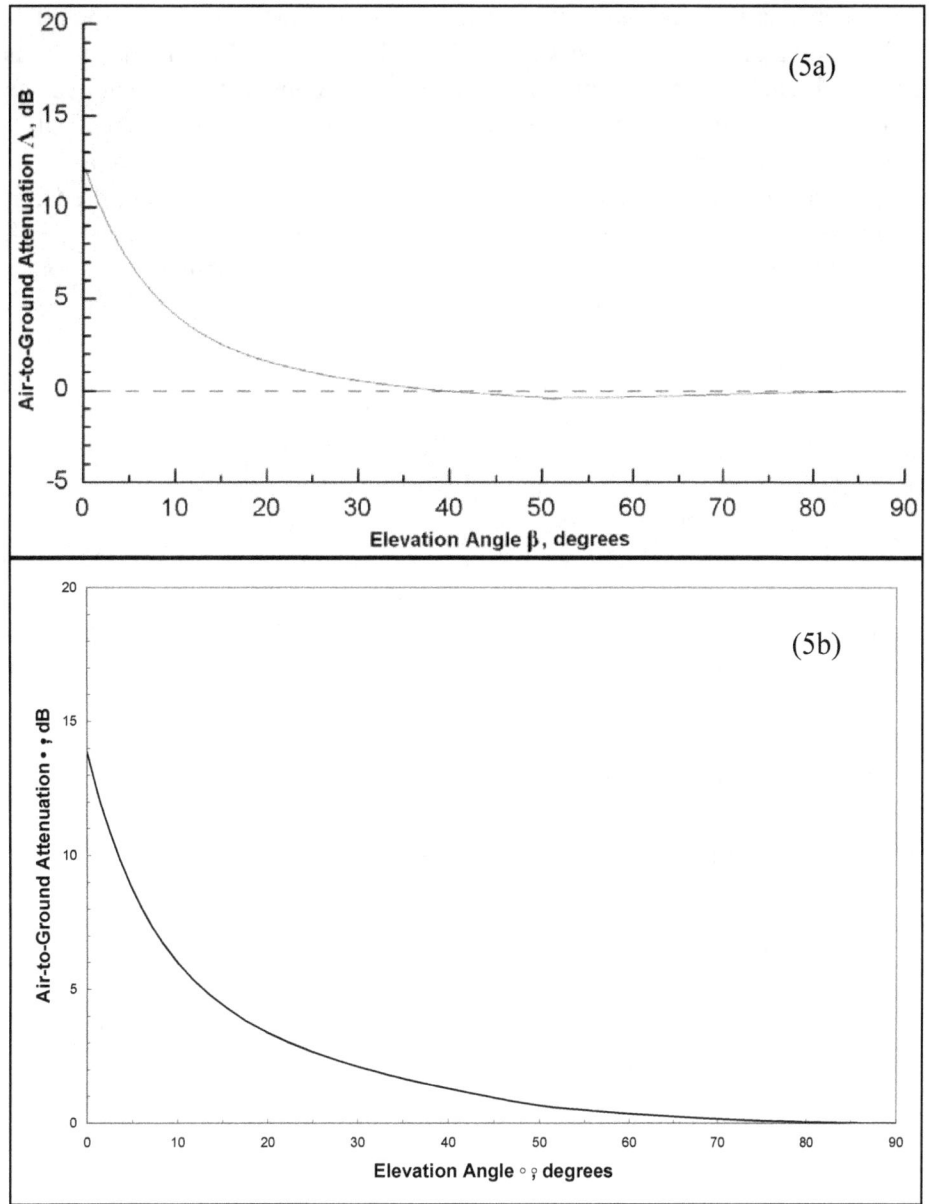

Figure 5. Lateral Directivity in INM for Jet Aircraft

In NMSim, lateral effects are basically synonymous with ground effect, since the model includes source directivity directly in its database – See Section 2.1. Ground effects in NMSim are based on the well-established work of Chien and Soroka, Embelton, Piercy and Daigle, as well as others [53-56]. This is the same physical model as contained in the revised SAE AIR 1751. The model takes into account the effects of varying ground surface, including both acoustically hard and soft surfaces, as well as a mix of both.

2.2.3 Terrain Shielding

In computing terrain shielding, the INM uses the well-established 'thin screen' equations of Maekawa [57], which have been adapted for use with earth berms and extensively validated by the Federal Highway Administration and others at least since 1978 [58]. An 18 dB attenuation cap is implemented in the INM terrain shielding formulation, as this value is considered a practical upper limit on shielding. Terrain shielding and lateral effects in INM are implemented as a logical "OR" function, i.e., both effects are computed and compared and the larger of the two effects are applied in the model. Although this approach does not consider interaction between the two phenomena, it allows for a seamless transition between the two. The approach has also proven quite successful in the FHWA modeling tools, and has been extensively validated to propagation distances of approximately 1,000 feet. Beyond 1,000 feet, refraction and scattering effects tend to dominate and a practical limit of between 18 and 25 dB can be expected. [65]

NMSim computes terrain effects using the Geometric Theory of Diffraction (GTD) presented by Rasmussen [33,34]. That method uses ray tracing for basic propagation, and various diffraction models for terrain effects. The diffraction model for ground effect is the finite ground impedance model noted above. Shielding is modeled by a general formulation for diffraction by a wedge of arbitrary angle. For the limiting case of a wedge with zero angle and hard surface, this reduces to the classic Maekawa thin screen case, but the GTD solution retains the interference pattern rather than following Maekawa's smoothed fit. Currently NMSim does not include an upper limit on terrain shielding. Simultaneous ground and diffracted paths are included in GTD, so the transition from shielded to unshielded (or combination of the two) is holistic, without switching. Rasmussen's algorithms were originally developed and validated for modest distances associated with highway noise, but were subsequently validated for distances of several kilometers [32].

2.3 Contouring/Grid Development and Noise Metrics

The underlying grids used in the two models for noise contouring are structured slightly differently. INM begins with a grid of regularly-spaced receptors, or grid points. It then recursively subdivides this grid of receptors, increasing receptor density in areas of high noise gradient, such as initial takeoff climb or where power setting changes occur. This approach ensures that any imprecision introduced by the interpolation associated with the noise contouring process is minimized. NMSim, on the other hand utilizes a regularly-spaced grid of receivers for contour generation. Both models use the NMPLOT software for the development of noise contours [60].

Tables 4 and 5 present the noise metrics that INM and NMSim, respectively, currently support. Highlighted in the tables are the metrics that have been identified in the GCNP MVS and ATMP processes as likely candidates for analysis. [66]

Table 4. Noise Metrics in INM

Noise Family	Metric Type	Noise Metric	Flight Multiplier			Averaging Time (hr)
			Day	Evening	Night	
A-Weighted	Exposure Based	SEL	1	1	1	--
		DNL	1	1	10	24
		CNEL	1	3	10	24
		LAEQ	1	1	1	24
		LAEQD	1	1	0	15
		LAEQN	0	0	1	9
		User-defined	A	B	C	T
	Maximum Level	LAMAX	1	1	1	--
		User-defined	A	B	C	--
	Time-Above Based	TALA	1	1	1	--
		User-defined	A	B	C	--
		%TALA	1	1	1	--
		User-defined	A	B	C	--
C-Weighted	Exposure Based	CEXP	1	1	1	--
		User-defined	A	B	C	T
	Maximum Level	LCMAX	1	1	1	--
		User-defined	A	B	C	--
	Time-Above Based	TALC	1	1	1	--
		User-defined	A	B	C	--
		%TALC	1	1	1	--
		User-defined	A	B	C	--
Tone-Corrected Perceived	Exposure Based	EPNL	1	1	1	--
		NEF	1	1	16.7	24
		WECPNL	1	3	10	24
		User-defined	A	B	C	T
	Maximum Level	PNLTM	1	1	1	--
		User-defined	A	B	C	--
	Time-Above Based	TAPNL	1	1	1	--
		User-defined	A	B	C	--
		%TAPNL	1	1	1	--
		User-defined	A	B	C	--
Time Audible	Time	TAUD	1	1	1	12
		%TAUD	1	1	1	12
		User-defined	A	B	C	T
		User-defined (%)	A	B	C	T
Change in Exposure	**Exposure Based**	DDOSE	1	1	1	12
		User-defined	A	B	C	T

Table 5. Noise Metrics in NMSim

Metric	Weighting(s)	Units
Lmax	A, C, Flat	dB
SEL	A, C, Flat	dB
Leq	A, C, Flat	dB
Ldn	A, C, Flat	dB
CNEL	A	dB
Time Above	A, C, Flat	Time
Time Above Ambient	A, C, Flat	Time or percent
Time Audible	d-prime	Time or percent
Spectral time history	1/3 octave bands	dB
Spectrum at receiver	1/3 octave bands	dB

2.4 Simulation vs. Integrated Models

The biggest single difference between NMSim and INM is that the former is a simulation model and the latter an integrated model. For the time audible calculation, both models must calculate (or reasonably represent) the time history of noise.

Consider an aircraft noise event that has a time history as sketched in Figure 6a. For the current study, the noise quantity for percent time audible is d-prime. NMSim, as a time step simulation model, will calculate a number of points on that time history, as sketched in Figure 6b. The time history is approximated by a series of trapezoids. INM, as an integrated model, calculates the noise at the closest point of approach (CPA) of each segment. That value is then assigned to the segment. For INM's usual application of computing DNL, the noise fraction is used to account for segment length and proper SEL contribution. For time audible, INM assigns the CPA noise value (d-prime) to the duration of the segment, i.e., the noise fraction is not used. This is equivalent to representing the time history by a series of rectangles, as sketched in Figures 6c and 6d. Figure 6c illustrates the segments using the same spacing as Figures 6a through 6b, and Figure 6d illustrates segments resulting from finer segmentation.

Three details are apparent from Figure 6:

1. Both simulated and integrated models can accurately approximate time histories if segmentation is fine enough.

2. The error of an integrated model will always be to overpredict (i.e., it will provide a conservative assessment of impact) while the error of a simulation model can be to over or underpredict, depending on the relation between the trapezoids and concave/convex portions of the time history.

3. Because trapezoidal representation of a curve is higher order than rectangular representation, a given degree of precision typically requires that an integrated model have finer segments than a simulation model.

2.5 Calculating Audibility

For the purposes of this study, audibility is defined as the ability for an attentive listener to hear aircraft noise. Based on signal detection theory[3,4], audibility depends on both the aircraft sound level ("signal") and the ambient sound level (background or "noise"). As such, true audibility is based on many factors, including the listening environment in which one is located. Detectability levels (d') calculated in support of this project are based on the signal-to-noise ratio within one-third octave-band spectra, using a 10log(d') value of 7 dB. Both INM and NMSim utilize d' to calculate audibility.

During initial GCNP noise modeling (pre the 1999 GCNP MVS), time above (TA) was used for impact analysis. Subsequent modeling has utilized time audible. Recognizing that the human auditory system itself has a noise floor, means to take this into account are required in the modeling of audibility. During the GCNP MVS, detailed one-third octave band and acoustic state log time history data were collected. Accordingly, site- and time-specific ambient sound level data were available for noise modeling. These data were compiled and mathematically combined with human auditory noise for modeling purposes.

Because these site-specific data are not available for the entire GCNP, NPS has identified ambient zones for park-wide contour analyses. Spectra are assigned to these large area zones for modeling.

Both the INM and NMSim audibility algorithm have been further developed since the GCNP MVS. In fact, the most recent developments have occurred since the October 2004 FICAN meeting. A complete history of the use of ambient and related methodologies in the two models is included in Appendix H. Currently the models are consistent in the evaluation of audibility for given signal and noise one-third octave band spectra. The process utilized herein by both models, which has been peer-reviewed by technical experts outside of the current FICAN project team, is considered to be the current state-of-the-art.

2.6 Other Capabilities

It is expected that time-based metrics will play a prominent role in conducting noise modeling studies in the National Parks, particularly the time audible noise metric. Traditional noise modeling studies compute noise contributions from each discrete event and simply aggregate the noise from all aircraft events within a particular time period. For the more traditional studies, which utilize noise-related descriptors, this aggregation of noise energy is appropriate. For studies involving time-based descriptors, it may be important to account for the effects of events, which overlap, in audible time, i.e., multiple events that can be heard simultaneously. This is certainly a concern for Grand Canyon National Park, which is subject to a substantial number of daily tour flights, particularly in the summer. However, it is not clear that this is an issue at any other parks in the U.S.

Both models have the capability to account for aircraft events that overlap in time. INM uses a measurement-based, empirical adjustment factor developed for the NPS based on GCNP operations

---

[3] Green, David M. and Swets, John A., "Signal Detection Theory and Psychophysics." New York: John Wiley and Sons, Inc, 1966.

[4] Fidell, et. al., "Predicting Annoyance from Detectability of Low-Level Sounds," Journal of the Acoustical Society of America, 66(5), November 1979, p. 1427 – 1434.

[1]. The algorithm may or may not be appropriate for use in other parks. If overlapping events were deemed an issue in other parks, a park-specific adjustment factor might have to be developed. Since NMSim is a time-based simulation model, it is capable of computing time overlap, if detailed flight schedule data are available for a particular study. The FAA, in cooperation with the air tour operators, has assembled a detailed flight schedule database for GCNP. This schedule data can be incorporated into a NMSim analysis. It is important to note that, based on available information collected to date under the ATMP effort (approximately 20 parks to date), such detailed schedule data do not exist for any other National Park.

In addition to its core noise computational module, the INM also contains a comprehensive aircraft performance model, which is based on the algorithms contained in SAE AIR 1845. This module supports detailed modeling of aircraft flight performance in the takeoff, low-altitude level flight (as is the case for tour aircraft operating in the National Parks) and approach regimes. Although NMSim does not contain a comparable capability, it has been shown in the current study that NMSim can be set up to efficiently use output data from the INM performance model.

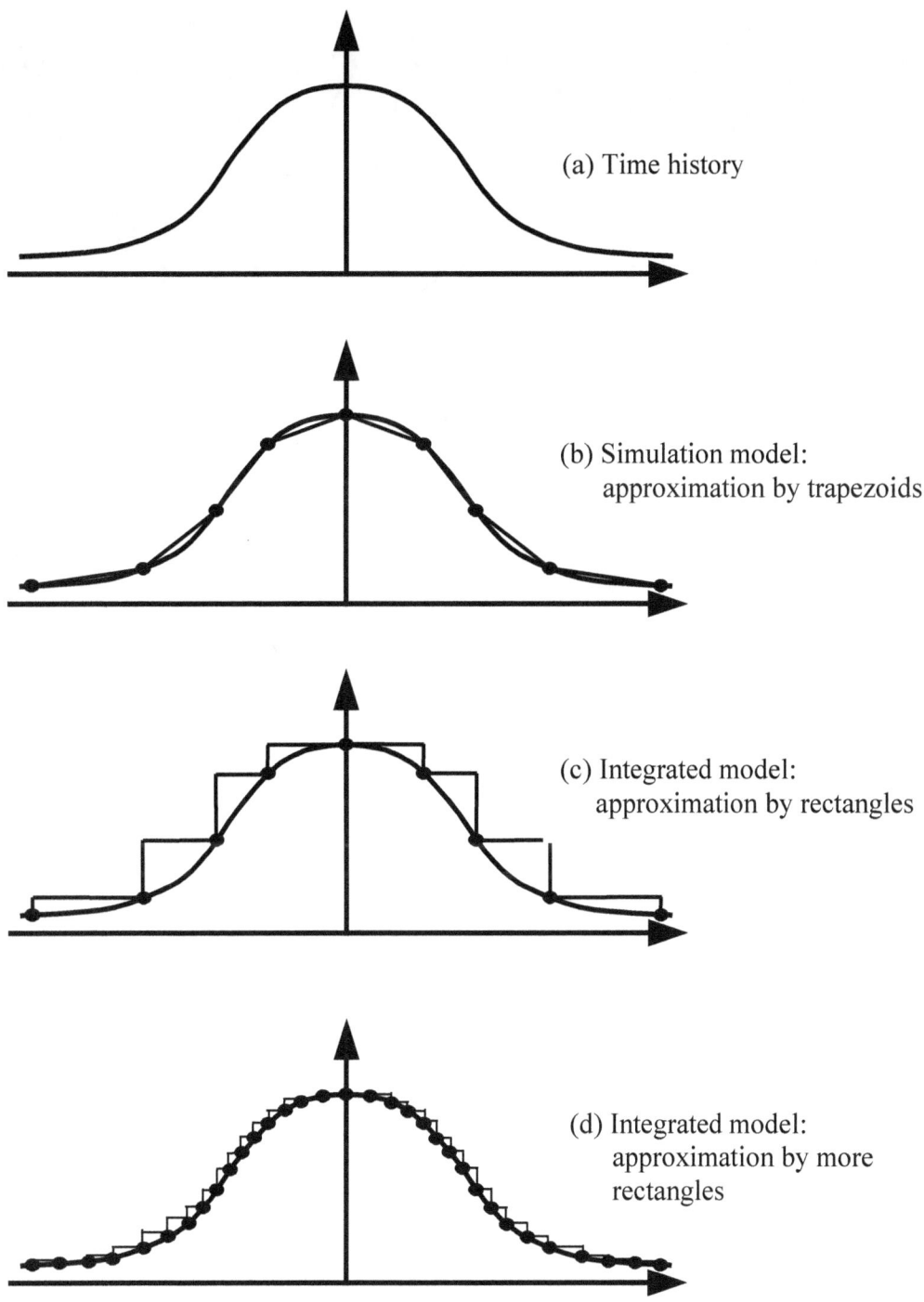

Figure 6. Approximation of Noise Level Time History By Simulation and Integrated Noise Models

3. Comparison of Model Calculations

As discussed in Section 2, although the two models both rely on sound fundamental physics, they generally use different formulations to account for the same propagation phenomenon. The purpose of this section is to provide the reader with a general sense for the effect these differing formulations have on the computed noise. Section 3.1 presents model comparisons for some fairly simplistic/generic cases. Section 3.2 presents some model comparisons specific to the GCNP MVS, which has become a de facto standard test case for this study. Section 3.3 presents a statistical assessment of the two models' performance based on the "gold standard" data collected in the GCNP MVS.

3.1 Generic Parametric Studies

Comparison of Source Data: Since the two models describe the source in fundamentally different ways, it was necessary to "translate" the noise level data from one model into a format, which is compatible with the other model. This translation would allow for a direct "apples-to-apples" comparison. The easiest way to accomplish this was to run a series of level flyovers in NMSim and generate INM NPD curves for a few of the more common aircraft in the database of the two models. Figures 7 and 8 present a comparison of the NPD data generated by NMSim with that used in the INM. As can be seen, while most SEL values are within approximately 2 dB of each other (Figure 7), some L_{ASmx} values (most notably the DHC6QP) are about 6 dB different (Figure 8). These differences are considered reasonable. The larger differences for L_{ASmx} are due to the directivity of the sources, a factor included in NMSim, and are not unexpected. Most notably, these differences have been substantially reduced when compared with the October 22, 2004, version of this document. These comparative improvements relating to the October version can be attributed to: (1) elimination of erroneous input source data in the NMSim hemisphere development process; (2) re-averaging of individual input events in the NMSim hemisphere development process; and (3) accounting for the differing atmospheric absorption algorithms in the two models (see Figure 4).

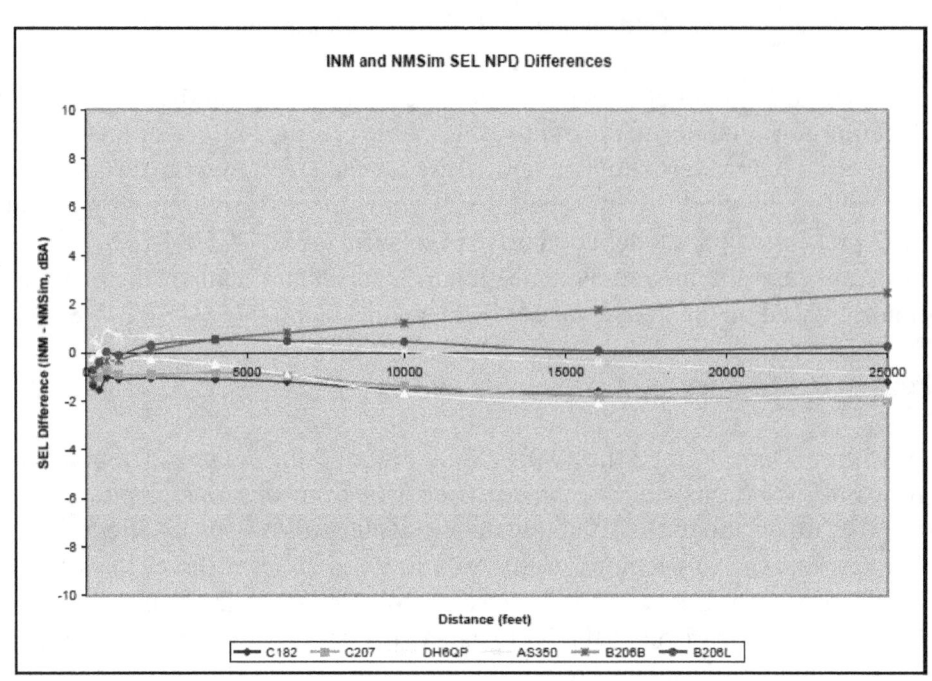

Figure 7. Comparison of INM and NMSim SEL NPD Data

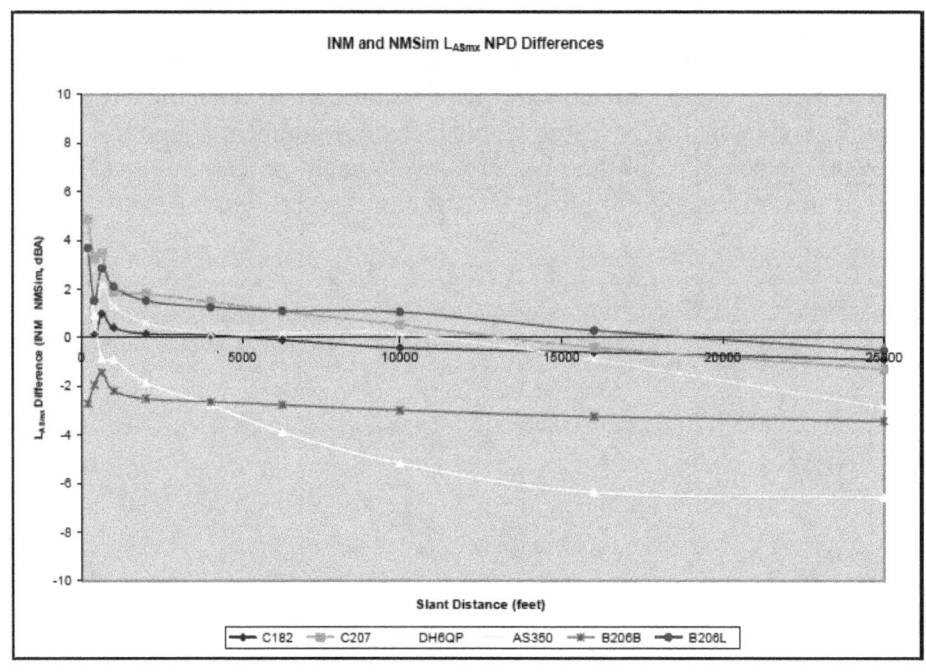

Figure 8. Comparison of INM and NMSim L_{ASmx} NPD Data

Atmospheric Absorption: Figure 9 presents several noise level differences as a function of distance, computed by subtracting levels computed using ARP 866a absorption and ISO 9613 absorption for several combinations of atmospheric conditions (77 degrees Fahrenheit / 70 percent relative humidity, 85/35, 85/85, 40/85, and 40/55). As can be seen, on average the differences are generally less than 1 dB – similar to Figure 4.

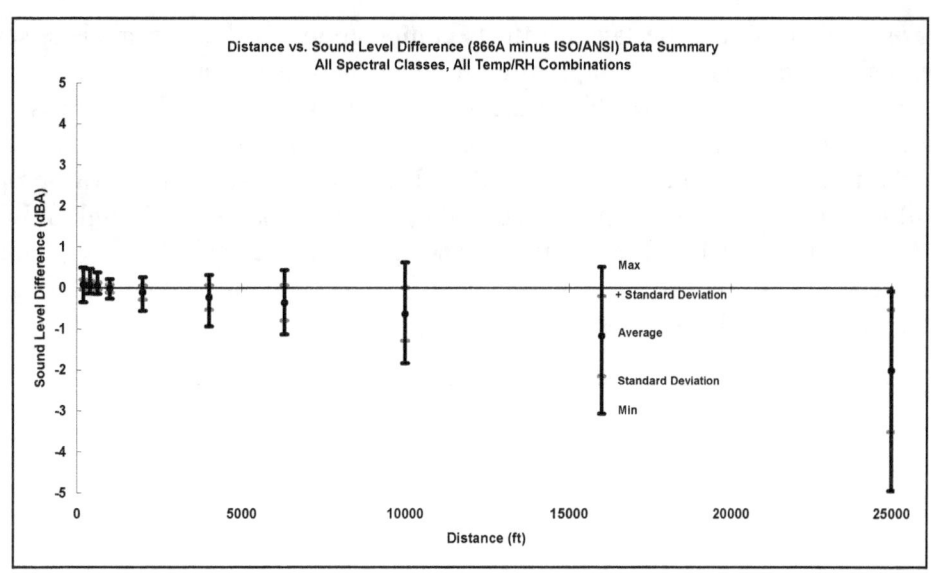

Figure 9. Comparison of INM and NMSim Atmospheric Absorption Effects on Noise Data

Lateral Effects: To examine the lateral effects computations in the two models, sensitivity tests were conducted using data for the DHC6QP aircraft. In this case, a 1,000-ft straight, level flyover was run at constant speed and power setting, with receptors positioned along a line perpendicular to the flight track, beginning directly below the track and extending out to 25,000 ft in 500-ft increments – with propagation over acoustically soft ground. Figure 10 shows the difference in the L_{ASmx} computed by the two models as a function of distance. In general, the differences are fairly small.

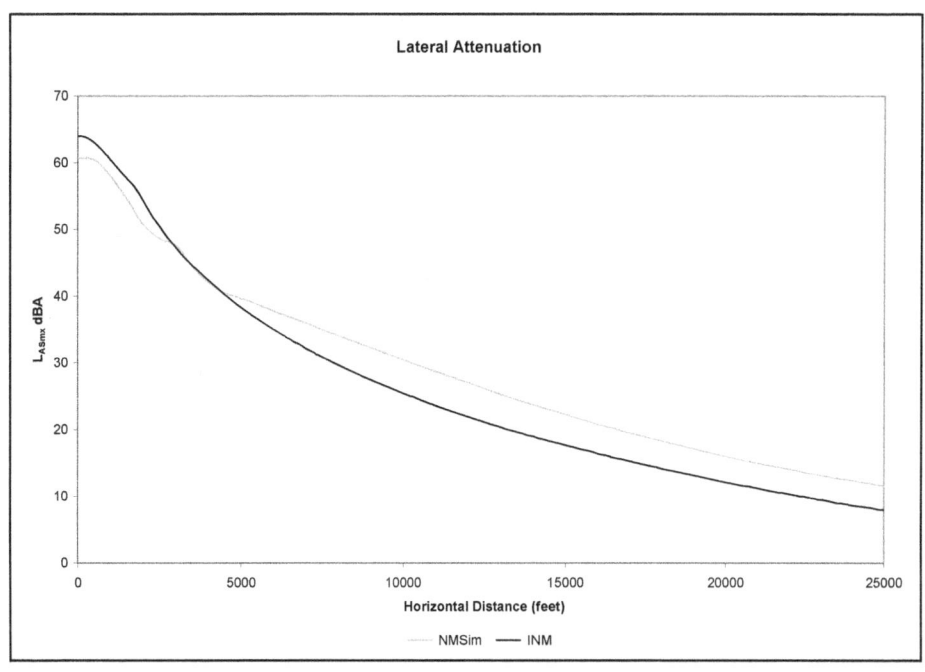

Figure 10. Comparison of INM and NMSim Lateral Effects

Terrain Shielding: To examine the lateral effects computations of the two models, sensitivity tests were conducted using data for the DHC6QP aircraft. The analysis conducted herein is similar to that conducted for lateral effects, above. In this case, a 1,000-ft straight, level flyover was run at constant speed and power setting, with receptors setup along a line perpendicular to the flight track, beginning directly below the track and extending out to 25,000 ft in 500-ft increments – with propagation over acoustically soft ground. The primary difference in this case was that a 500 ft high infinitely long hill was introduced at a distance of 1,250 ft. Figure 11 shows the difference in the L_{ASmx} computed by the two models as a function of distance. In general, the differences are fairly small, although the effect of capping terrain shielding in INM is evident.

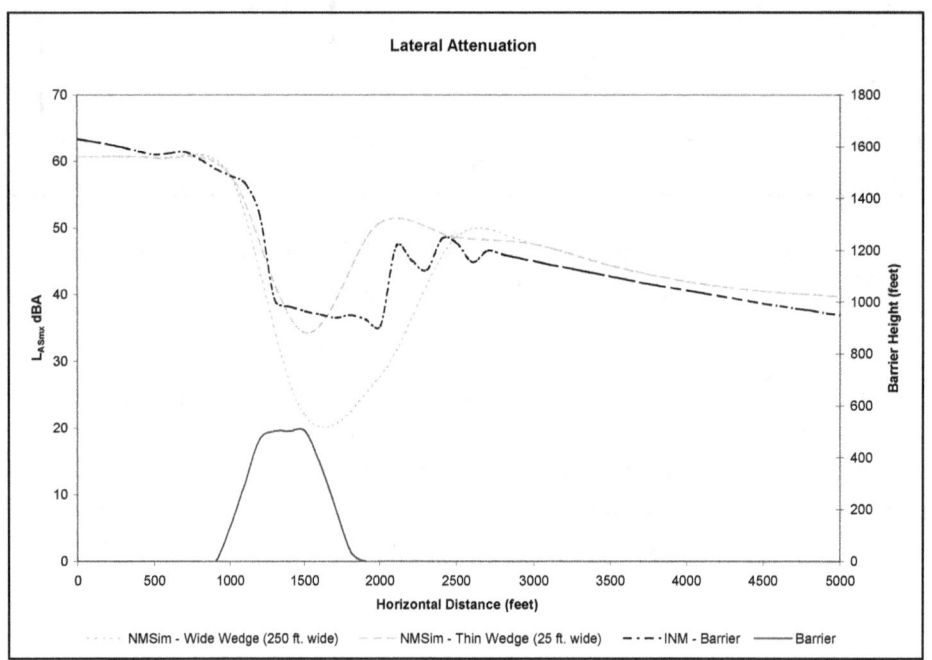

Figure 11. Comparison of INM and NMSim Terrain Shielding

Contouring: Both INM and NMSim use the NMPlot noise contouring software to generate noise-related contours from a set of input grid points. However, NMPlot has a series of user-selectable options, which can result in slightly different contours being generated from the same input grid. To confirm that this was not an issue in the current analysis, a common noise grid for Aircraft Scenario 4 (see Section 4.1.4) was separately input to NMPlot as configured in INM (with its default settings) and as configured in NMSim. The output contours are overlaid and shown in Figure 12

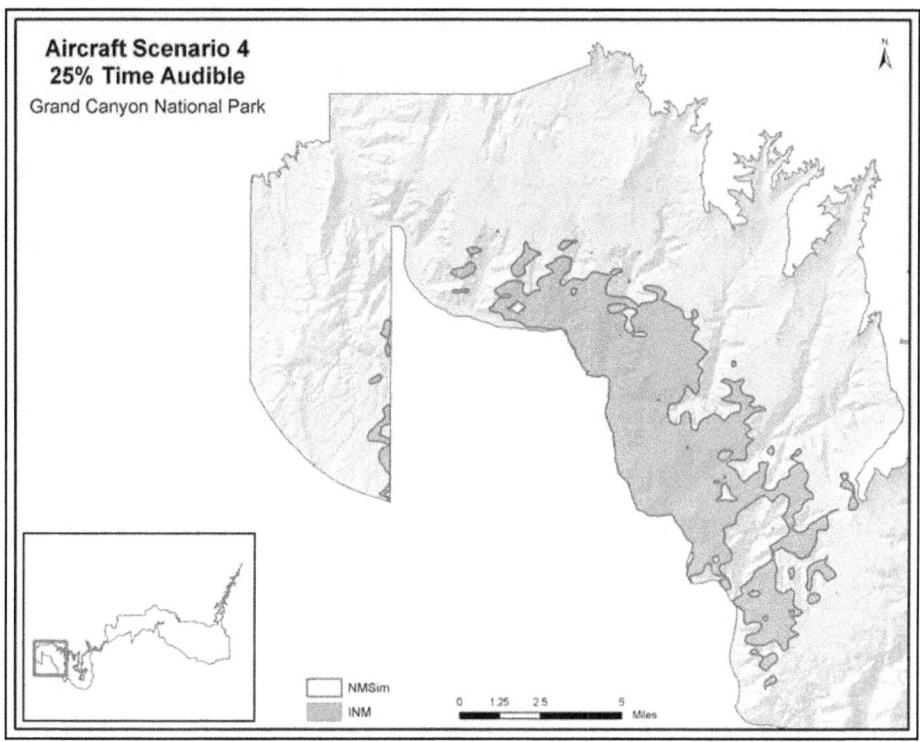

Figure 12. Comparison of INM and NMSim Contouring

3.2 Grand Canyon Noise Model Validation Study (GCNP MVS)

It is clear from the results presented in Sections 3.1 and 3.2 that INM and NMSim are providing similar output, on average. This section compares the output of the two models to the measured time audible data collected in the GCNP MVS – the so-called "gold standard" dataset for assessing model performance. Included in this section are the measured versus modeled graphics, as well as a statistical assessment of each model's performance relative to the gold standard dataset.

Figure 13 presents a side-by-side comparison of measured percent time audible (%TAud) data from the GCNP MVS modeled using INM 6.2 and NMSim. Also shown in the graphic is a linear regression through both data sets along with the perfect agreement line. Figure 13 can be directly compared with Figure 12 from the GCNP MVS, which is reproduced herein as Figure 14 for comparative purposes. As can be seen, both models are performing well, on average – not a surprise, given the results presented in the two previous sections.

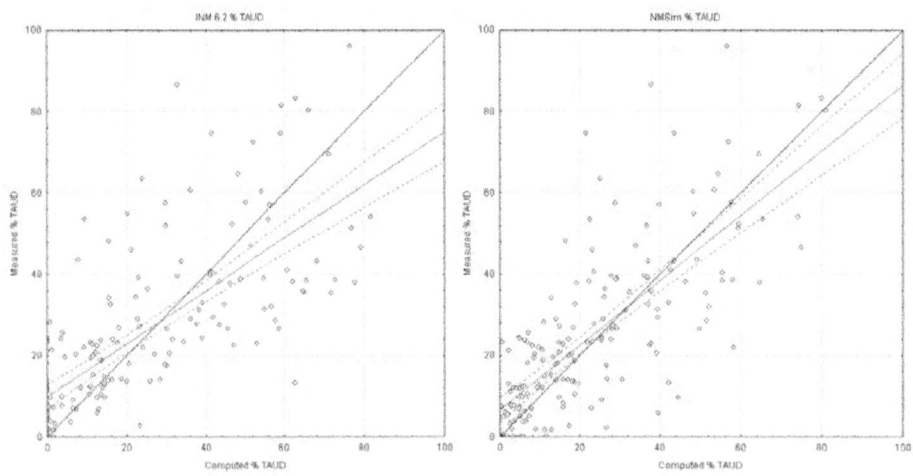

Figure 13. INM 6.2 and NMSim Modeled vs. Measured %TAud

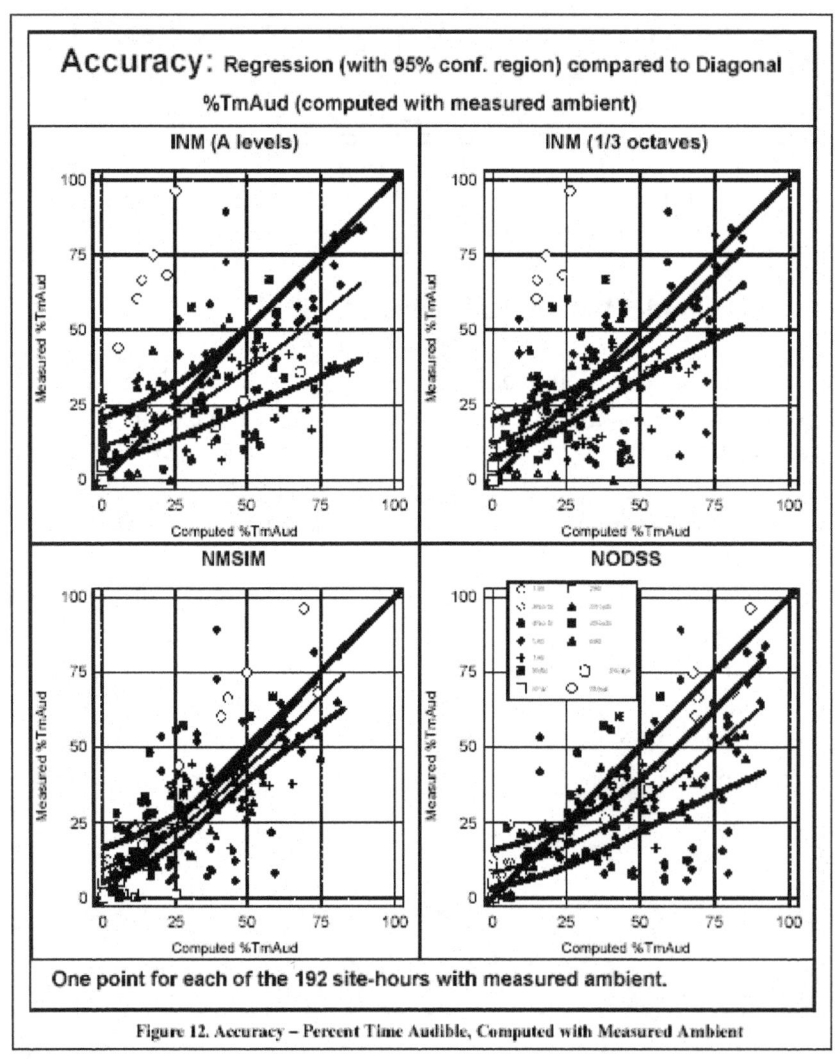

Figure 14. Grand Canyon Noise Model Validation Study Figure 12

Table 6 presents a quantitative statistical comparison of the data presented in this section. As was done in the MVS, the following statistical measures are included for each model: overall error, bias, random error and correlation coefficient. Each of these measures is shown separately for the individual measurement periods at a site, as well as aggregated for each site. Appendix G presents a description the statistical measures used. Not surprisingly, the statistical measures for the two models are very similar. INM 6.2 performs slightly better than NMSim in terms of bias; and NMSim performs slightly better than INM6.2 in terms of scatter, as measured by the error and confidence interval metrics.

Table 6. $\%T_{Aud}$ Statistics

	Individual Hours					Site Groups				
	Overall Error	Bias	CI	Random Error (log)	Corr. Coeff.	Overall Error	Bias	CI	Random Error (log)	Corr. Coeff.
INM 6.2	16.4	-1.8	2.4	15.1	0.73	12.2	-1.4	6.2	11.2	0.81
NMSim	13.0	-3.2	1.9	13.2	0.81	8.9	-3.7	4.1	4.1	0.89

Figures 15 and 16 present a comparison of the bias and confidence intervals derived for NMSim and INM 6.2. These figures present the results for all data. In three of four cases, INM 6.2 and NMSim 95% CIs encompass the 0% TAud lines, indicating that there is 95% confidence that there is no difference between the measured and modeled data from the two models. For individual hours, the 95% CI for NMSim does not encompass the zero line.

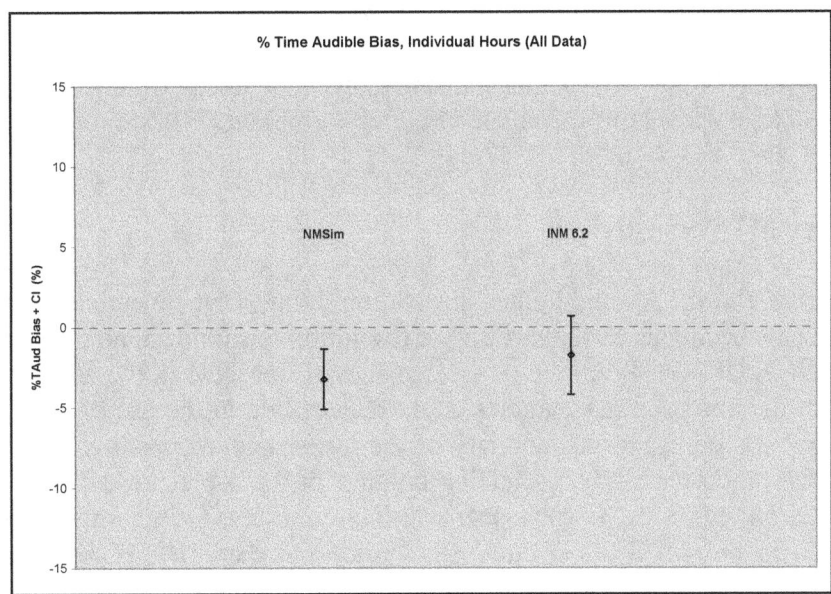

Figure 15. $\%T_{Aud}$ Model Bias and CIs, Individual Hours

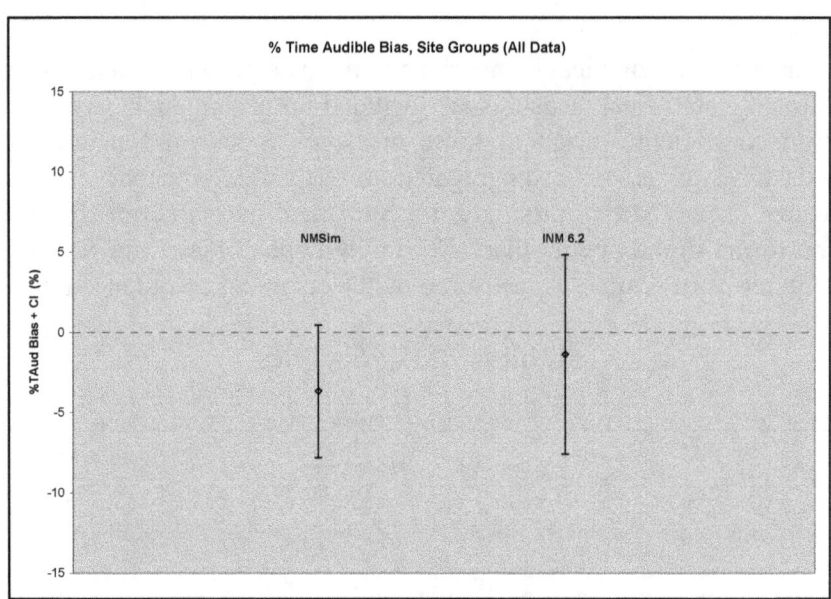

Figure 16. %T$_{Aud}$ Model Bias and CIs, Site Groups

INM Segmentation: As is discussed further in Section 4, model outputs have shown to be dependent on the resolution of data used to define aircraft position in the two models (i.e., segmentation in INM and time step in NMSim). To explicitly check sensitivity with respect to this variable, the INM flat earth model validation case was exercised with a total of 999 points per flight track. Note that this is as compared to other sensitivity model runs which utilized 300 track points. The maximum change in %TAUD any of the GCNMV data points was -1.9 (from 68.0% to 66.1% TAUD). The average change for all data was -0.5%. 999 points translates into a segment distance of up to 630 feet; an aircraft traveling 100 knots covers this distance in under 4 seconds. NMSim utilized a 10-second time step for these model runs.

3.2.1 Grand Canyon MVS Sensitivities

Elevation and Source Data: As highlighted in Section 1, the first objective of this study was the assessment of INM enhancements to support higher resolution terrain data, an obvious improvement in an environment such as GCNP. Preliminary assessments of the INM in the MVS seemed to indicate that at least some of the observed inaccuracies could be ascribed to the use of the lower fidelity 3CD terrain data. Table 7 is a comparative summary of the terrain elevations at each of the measurement sites from the model validation study. The first column in the table presents the measurement site location from the GCNP MVS. The subsequent six columns present the site altitudes from various sources: as reported in the MVS, 3CD data, 30m GridFloat data, 10m GridFloat data, DLG data directed converted to GridFloat data using the GlobalMapper software, and DLG data from NMSim, respectively. The last five columns present differences between the various terrain data sources and those presented in the GCNP MVS. The differences are generally within about +/- 100 ft, but in most cases appear to be somewhat random, at least for the DLG, and grid float data. It is clear that the 3CD data are of relatively poor quality.

Table 7. Comparison of Terrain Data Elevations

Location	GCNP MVS	3CD	GridFloat: LowRes	GridFloat: HighRes	DLG-> GridFloat	NMSim DLG	GCNP MVS - 3CD	GCNP MVS - GridFloat LowRes	GCNP MVS - GridFloat HighRes	GCNP MVS - DLG-> GridFloat	GCNP MVS - NMSim DLG
1A	3680	3612	3690	3685	3637	3719	68	-10	-5	43	-39
1B	3640	3624	3641	3648	3609	3674	16	-1	-8	31	-34
2A	3810	3630	3815	3812	3687	3783	180	-5	-2	123	27
2B	3660	3645	3687	3675	3681	3724	15	-27	-15	-21	-64
2C	3750	3595	3745	3755	3672	3802	155	5	-5	78	-52
2D	3720	3341	3729	3728	3615	3751	379	-9	-8	105	-31
3A	3650	3411	3743	3783	3144	3795	239	-93	-133	506	-145
3B	3560	3213	3613	3673	3218	3409	347	-53	-113	342	151
3D	3580	3378	3704	3672	3469	3604	202	-124	-92	111	-24
3H	4110	3792	4132	4154	3915	4182	318	-22	-44	196	-72
3J	4010	3727	4023	4014	3871	3908	283	-13	-4	139	102
3K	3630	3411	3684	3678	3603	3681	219	-54	-48	27	-51
4A	4870	4738	4895	4900	4757	4902	132	-25	-30	113	-32
4B	4890	4502	4921	4916	4757	4933	388	-31	-26	133	-43
4C	4900	4799	4917	4910	4793	4890	101	-17	-10	107	11
4D	4820	4809	4887	4893	4757	4877	11	-67	-73	63	-57
4E	5140	4997	5105	5136	5043	5091	144	35	4	97	49
5A	7960	5666	7952	7968	7874	7925	2295	8	-8	86	35
5B	8040	5859	8007	8046	7971	8039	2181	33	-6	69	1
6A	7210	7198	7205	7206	7218	7180	12	5	5	-8	30
6C	7240	7198	7252	7251	7218	7246	42	-12	-11	22	-6
6D	7290	7198	7297	7293	7230	7284	92	-7	-3	60	6
7A	4270	4167	4258	4335	4194	3979	103	12	-65	76	291
7B	5570	5002	5562	5587	5549	5669	569	9	-17	21	-99
7C	5530	5572	5560	5578	5522	5514	-42	-30	-48	9	16
7E	3970	3460	4013	4047	3936	4033	510	-43	-77	34	-63
7G	5370	4953	5402	5404	5391	5592	417	-32	-34	-21	-222
7H	5620	4718	5602	5619	5552	5607	902	18	1	68	13
8A	7010	6581	6977	7015	6910	6953	429	33	-5	100	57
8B	6760	6623	6755	6765	6713	6512	138	5	-5	47	248
8C	7010	6743	6974	6992	6946	7025	267	36	18	64	-15
8D	6940	6794	6952	6966	6890	6751	147	-12	-26	50	189
8E	6940	6797	6955	6958	6890	6948	143	-15	-18	50	-8
9A	6930	6149	6130	6130	6167	6125	781	800	800	763	805
9B	6920	5963	5969	5969	5911	5967	957	951	951	1009	953
9C	6910	5997	6044	6040	5729	5299	913	866	871	1181	1611
9D	6920	5969	5950	5949	5922	5943	951	970	971	999	977
9E	6920	5968	5956	5954	5919	5948	952	964	967	1001	972
9F	6910	5997	6044	6040	5729	5299	913	866	871	1181	1611

Given the somewhat random differences seen in the various terrain data sets used by the two models, yet keeping in mind that similarities in the fundamental physics of the two models, one might expect that the outputs of INM and NMSim would be similar, on average, but there would likely be some random differences when comparing predicted output from the two models at individual points. Figure 17 depicts the time audible computed by the two models at each of the model validation sites, plotted against one another. Also shown are two linear regressions (INM versus NMSim and vice-versa). As can be seen, the agreement between the two models (on average) is quite good.

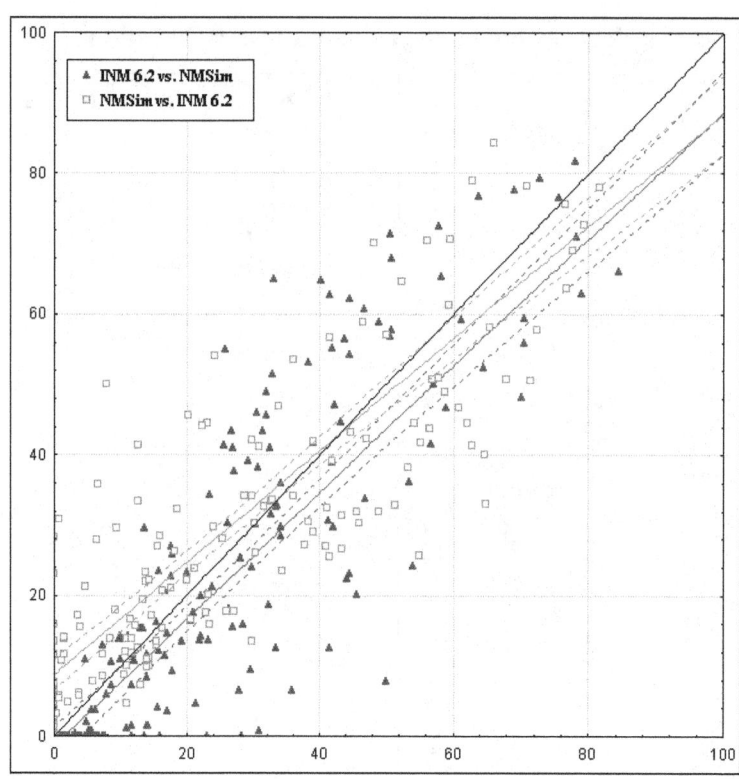

Figure 17. Comparison of INM and NMSim Time Audible With Terrain

Given the somewhat random elevation variations shown in Table 7, the time audible was recomputed in the two models at each of the model validation sites, but with the terrain capability in the two models not invoked, i.e., the models were run assuming a flat-earth model, referenced to the GCNP airport altitude of 6,600 ft. The results are shown in Figure 18. As can be seen, the two models agree well on average, and the scatter is reduced somewhat, indicating the overall sensitivity of model output to the terrain data used. However, there is a slight bias, indicating INM predicts higher audibility than NMSim. One other notable item in Figure 18 is that most of the 0% time audible computations observed in the October 2004 version of this report are no longer there. This can be attributed to the updated audibility algorithm.

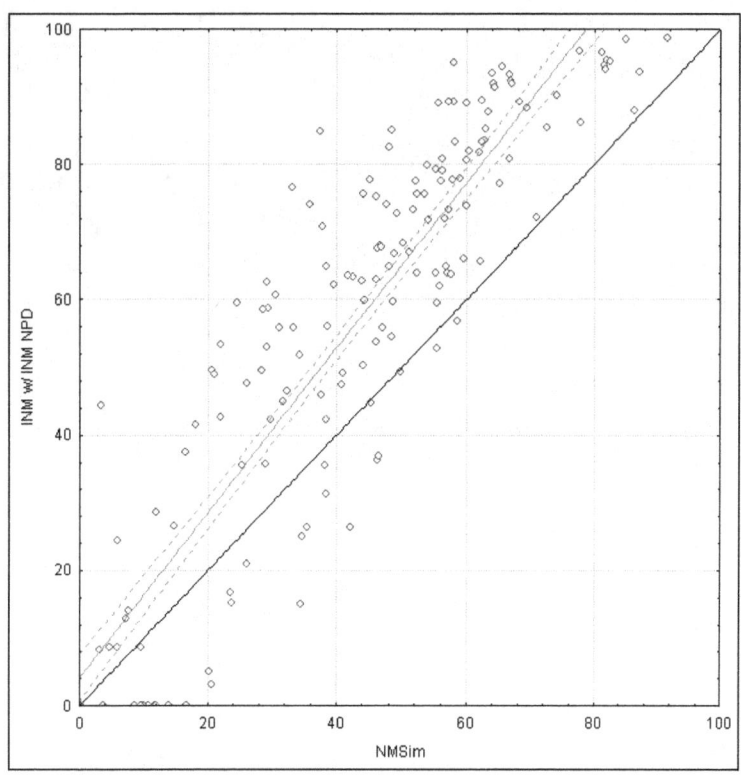

Figure 18. Comparison of INM and NMSim Time Audible
With Flat Earth

As discussed in Section 2.0, INM uses spectral classes while NMSim uses full spectral time histories. To ensure a more apples-to-apples comparison, Figure 18 was rerun in NMSim but using an omni-directional source spectrum, which more closely emulates the INM spectral class approach. The results are shown in Figure 19. As can be seen, the bias, although it still exists, has been noticeably reduced.

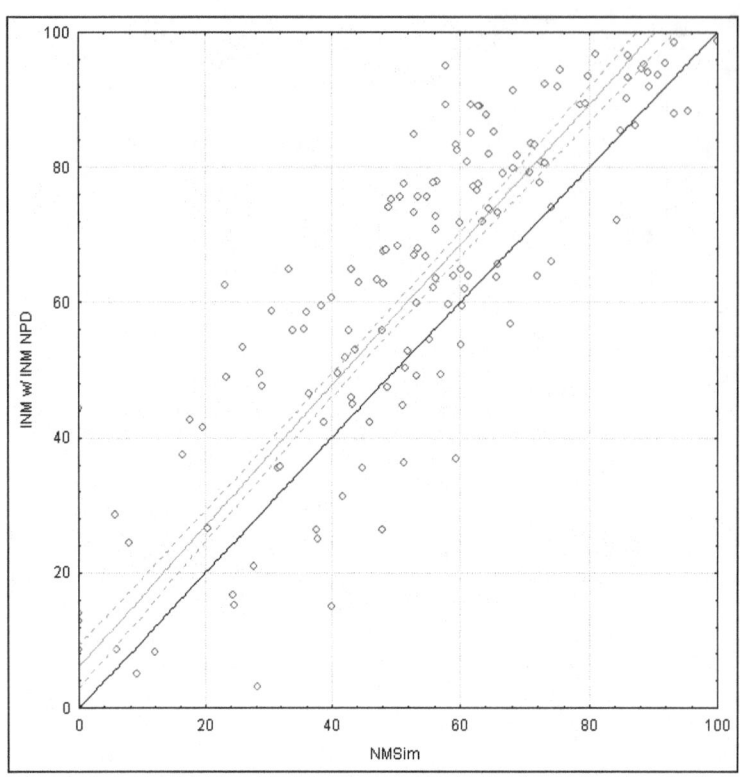

Figure 19. Comparison of INM and NMSim Time Audible
With Flat Earth and Identical NPDs

To further investigate the differences in INM and NMSim source data, data representative of single events were compared for each aircraft type. Specifically, computed audibility, in minutes, was compared for the two models; Figure 20 presents the results. For most aircraft, the agreement is generally good. The noticeable exception is the C207.

The agreement seen in Figures 19 and 20 shows that, for the same sources, propagation physics is consistent between both models. The source directivity included in NMSim calculations for Figure 18 is, however, real, and is readily seen in the original MVS source data recordings. Calculated time histories for individual events generally agreed around the closest point of approach, but at times well before or after CPA the levels from omnidirectional sources (as used in IMN) were higher than for directional sources (as used in NMSim). That is the reason for the offset seen in Figure 18. The offset does not appear in Figure 17, where NMSim uses directional sources but both models include terrain. Terrain shielding is increasingly likely to occur at longer distances, so the directivity effect may not be as important at Grand Canyon because those portions of the time history are blocked. Directivity may be a more important factor at parks that do not have substantial terrain shielding.

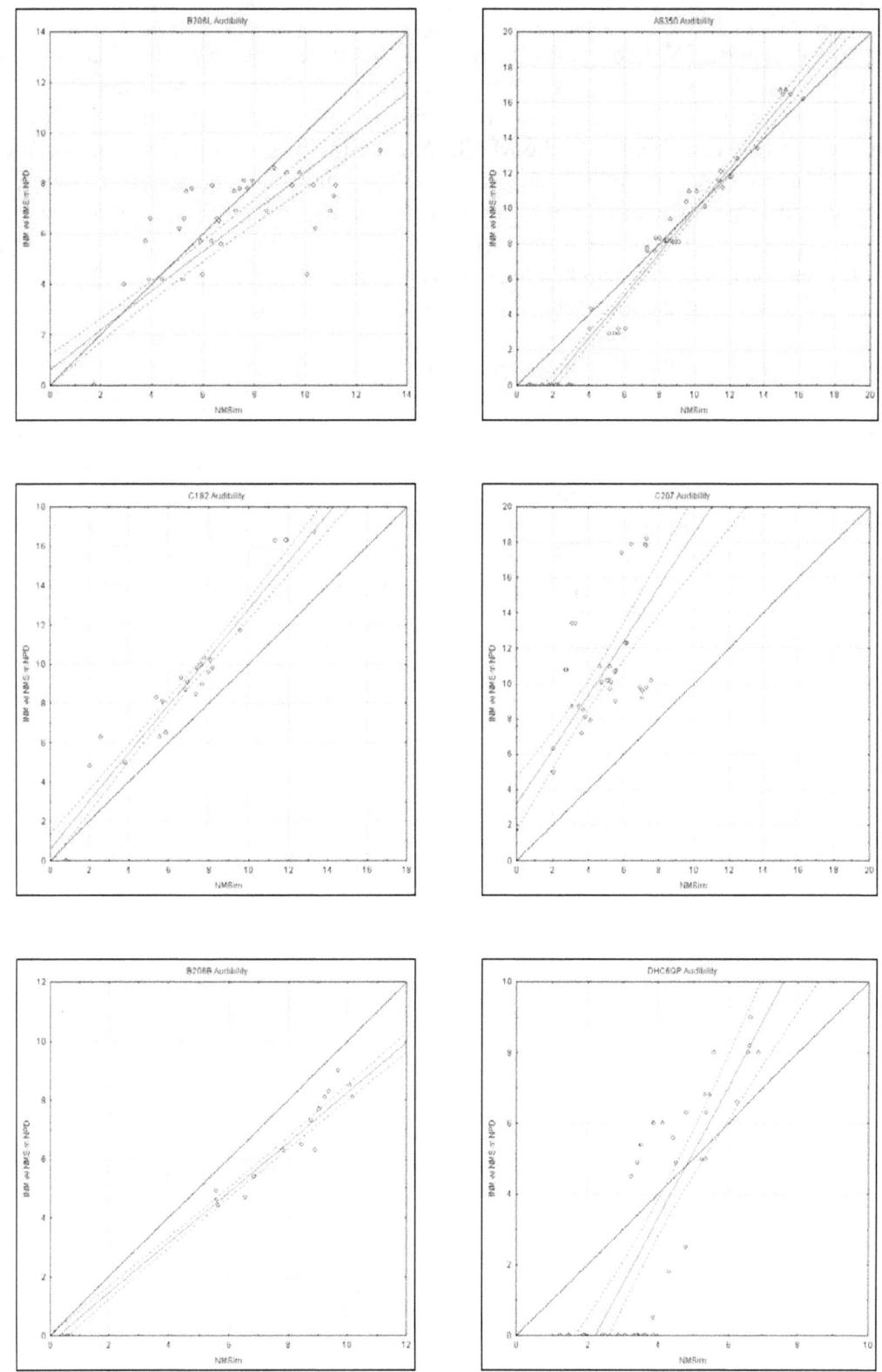

Figure 20. Single-Event Time Audible Comparisons

Overlapping Noise Events: INM computes the time audible for each event individually, then sums all events and divides by time to obtain %TAUD; this can yield %TAUD values greater than 100 percent. This is adjusted by an empirical compression algorithm. NMSim, when run with the scheduler, accounts for overlap between events. In the MVS, time audible was computed based on the summed sound levels, accounting for overlap. That procedure assumed that during overlap periods people would detect the combined noise of two or more aircraft. Unless two aircraft are of the same type and in the same position, however, it is likely that people would detect the louder of the two. The NMSim analysis was revised such that detectability during overlap periods could alternatively be based on the loudest aircraft. This lowered the total time audible by a modest amount.

Figure 21 compares the two NMSim overlap methods with the INM compression algorithm. Shown are calculated sum-of-audible time results from 0 to 140 percent, vs. scheduler (accounting for overlap) results from 0 to 100 percent. Part a shows scheduler results based on hearing the summed noise levels, and Part b shows scheduler results based on hearing the maximum of multiple events. A 1:1 line and the INM compression algorithm are shown on both plots. The INM algorithm agrees well with the scheduler results, and agrees better with the original NMSim (summing events) than with the alternative (hear the loudest event) method. Differences between the two methods are, however, modest, and overlap is not a dominant effect at lower %TAUD.

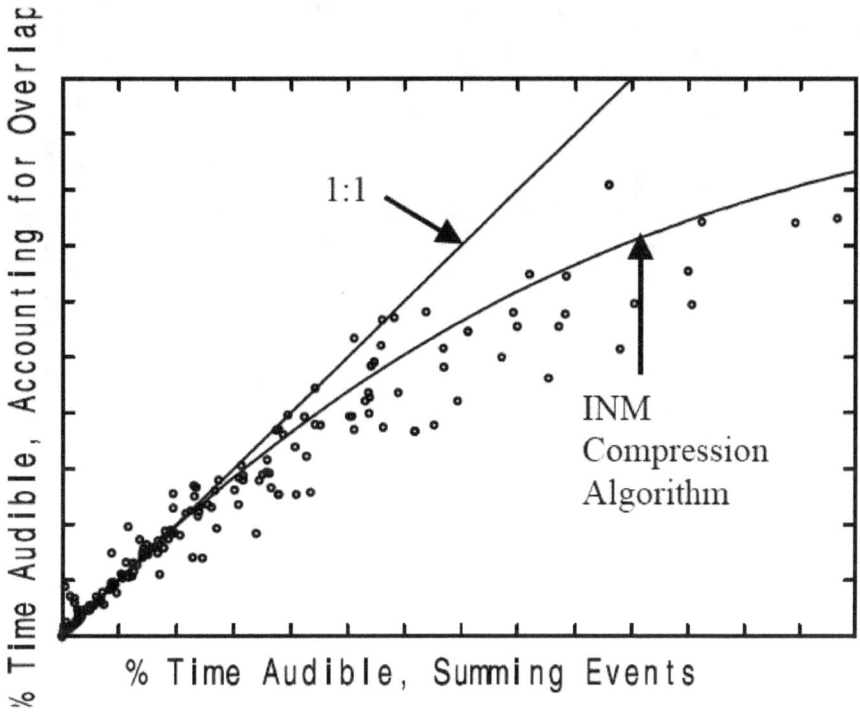

a. Audibility based on sum of levels of overlapping events

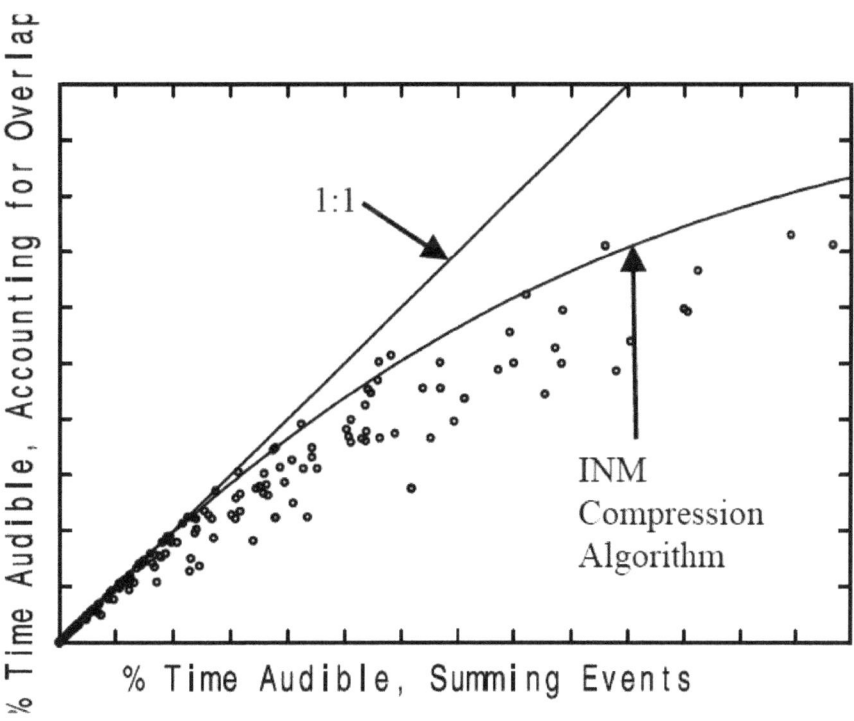

a. Audibility based on maximum level of overlapping events

Figure 21. Effect of Overlap on %TAUD
NMSim Scheduler Results Compared with INM Compression Algorithm

4. Grand Canyon Noise Analysis

The primary purpose of Section 4 is to present a comprehensive, comparative assessment of aircraft audibility in Grand Canyon National Park, using both INM Version 6.2 and NMSim. The analysis presented in this section supports: (1) the joint FAA/NPS Alternative Dispute Resolution (ADR) process in GCNP; and (2) from the standpoint of FICAN, the analysis serves as a de facto case study.

The input data used to support the analyses presented in this section come from several sources, including:

Flight Trajectories: Provided by the National Oceanographic and Atmospheric Administration (NOAA) mapping group within the FAA: (1) climb and descent profiles in the terminal area, and transitions from NOAA enroute altitudes, based on INM's performance model (commercial air tours and related flights); (2) FAA's Enhanced Traffic Management System (ETMS) (commercial overflights, GA and military only); and (3) FAA's PDARS system, as a verification of the ETMS data.

Operations (Types and Schedule): Based on: (1) a comprehensive spreadsheet maintained by the FAA with joint cooperation from the GCNP air tour operators; (2) ETMS (commercial overflights, GA and military only); and (3) PDARS as a verification of the ETMS data. For comparative purposes, data were drawn from the 3 operational sources for the same data. August 31, 2003, was selected because it reflected the average number of daily air tour operations over GCNP during a peak month in 2003.

Noise Source Data: From: (1) GCNP MVS (commercial air tours and related flights); (2) joint FAA/NPS field study (commercial overflights, GA and military only) – See Appendix G.

GCNP Ambient Sound Levels: Developed in support of previous FAA/NPS joint environmental analyses [61].

4.1 Contributions by Category

This section presents the computed time audible contours by operational category, computed by both INM and NMSim. Section 4.1.9 then presents the aggregate contributions from multiple operational categories. Since the purpose of this analysis is to compare the results generated by both INM and NMSim, the various aircraft flight scenarios are not directly attributed to specific flight operations around GCNP. Note, however, that they do represent operations for the complete fleet currently known to fly over various sections of the park.

4.1.1 Aircraft Scenario 1

Figures 22 and 23 depict the 25 % time audible contours associated with INM and NMSim, respectively, for Aircraft Scenario 1.

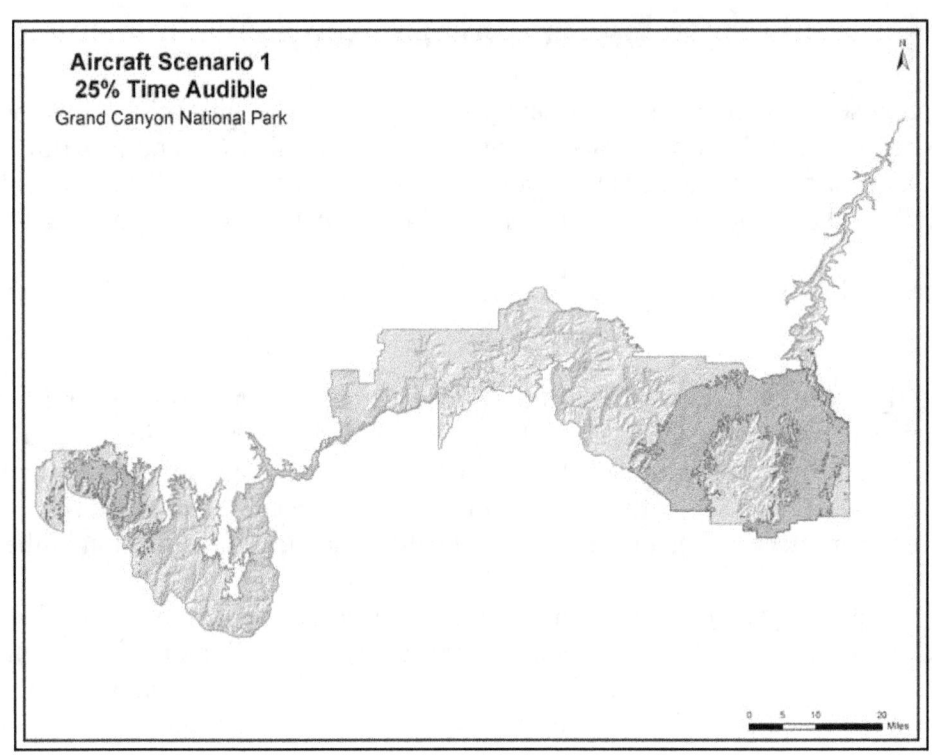

Figure 22. 25 %T$_{Aud,}$ Aircraft Scenario 1, INM
25% TAud = 23.9 % of Park

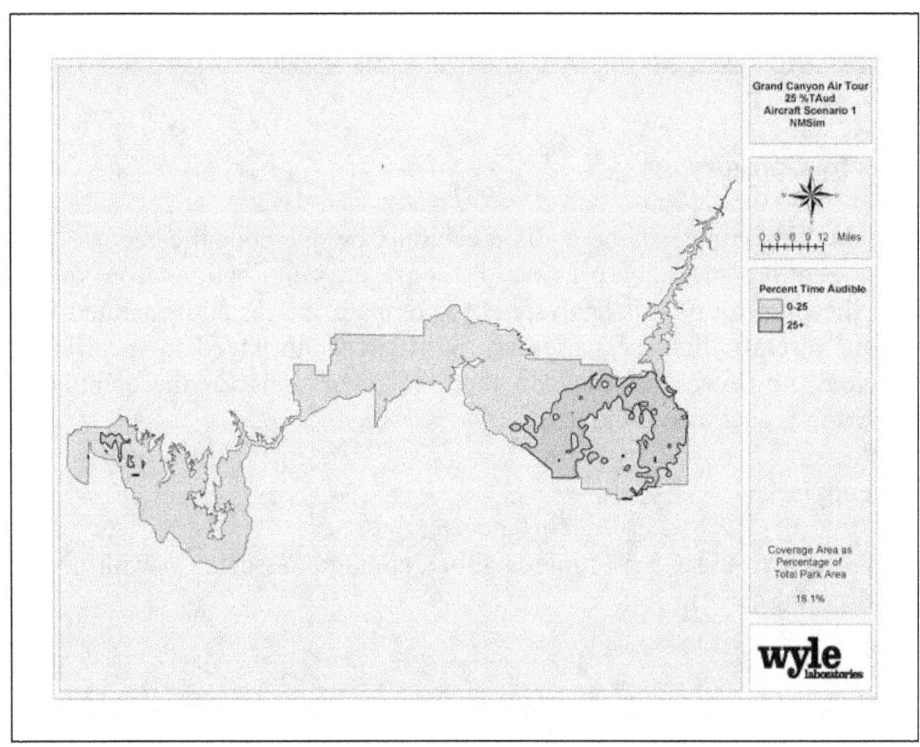

Figure 23. 25 %T$_{Aud,}$ Aircraft Scenario 1, NMSim
25% TAud = 18.1 % of Park

4.1.2 Aircraft Scenario 2

Figure 24 and 25 depict the 25 % time audible contours associated with INM and NMSim, respectively, for air tour related flights.

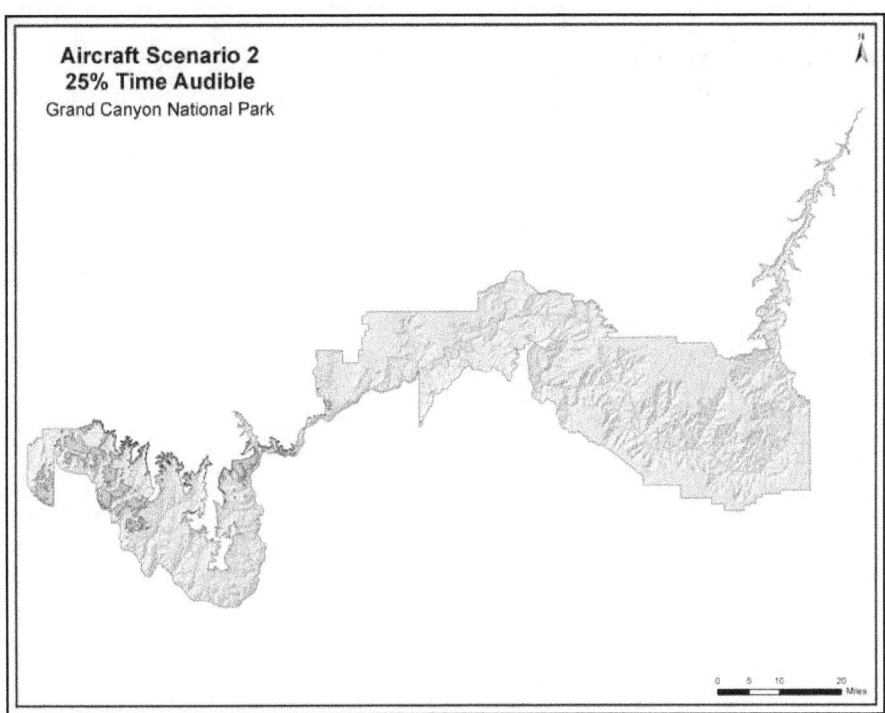

Figure 24. 25 %$T_{Aud,}$ Aircraft Scenario 2, INM
25% TAud = 7.7 % of Park

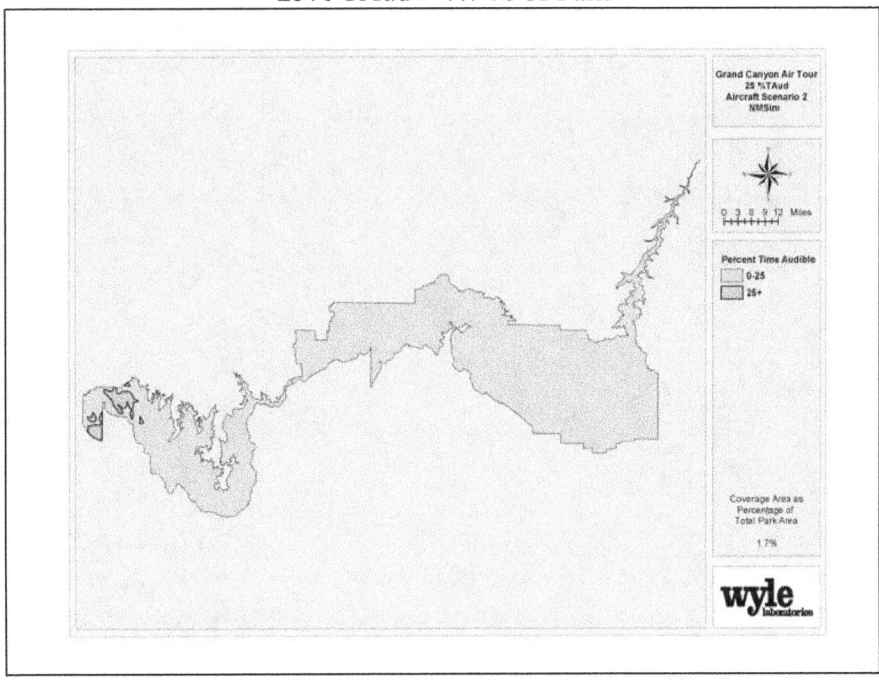

Figure 25. 25 %$T_{Aud,}$ Aircraft Scenario 2, NMSim
25% TAud = 1.7 % of Park

4.1.3 Aircraft Scenario 3

Figure 26 and 27 depict the 25 % time audible contours associated with INM and NMSim, respectively, for Hualapai exempted flights.

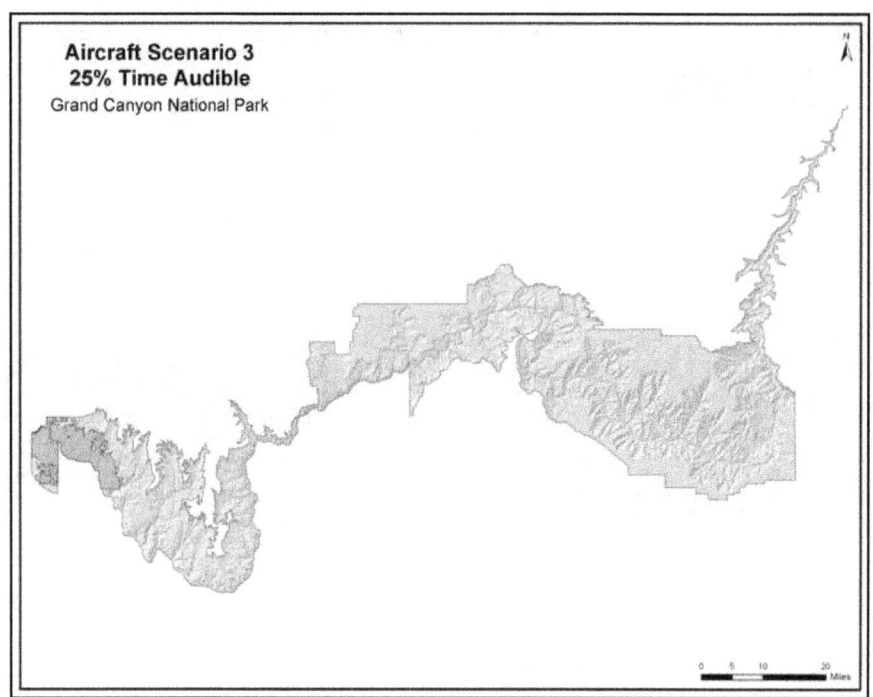

Figure 26. 25 %$T_{Aud,}$ Aircraft Scenario 3, INM
25% TAud = 4.2 % of Park

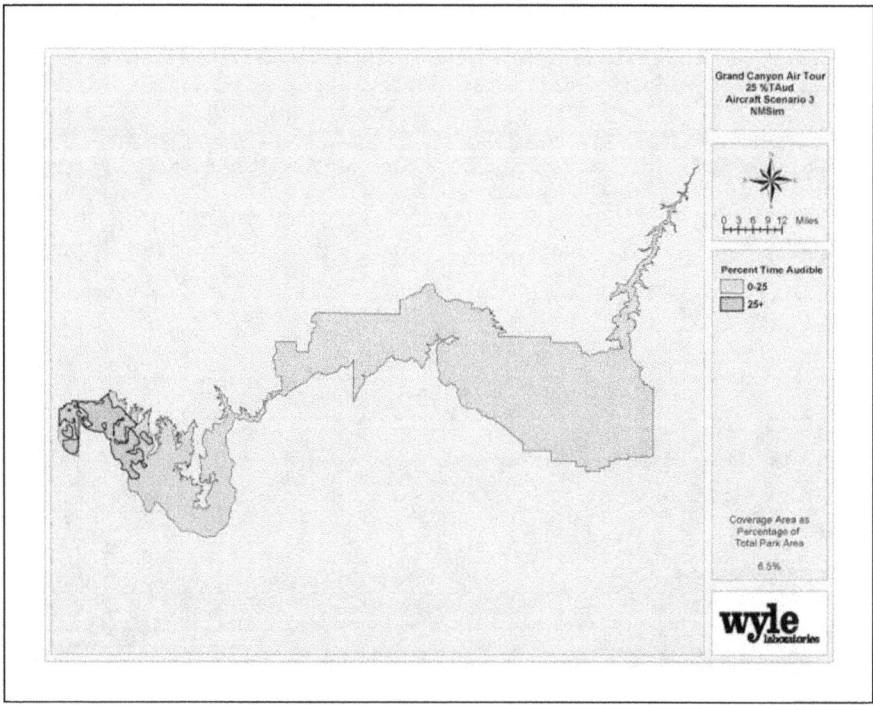

Figure 27. 25 %$T_{Aud,}$ Aircraft Scenario 3, NMSim
25% TAud = 6.5 % of Park

4.1.4 Aircraft Scenario 4

Figure 28 and 29 depict the 25 % time audible contours associated with INM and NMSim, respectively, for Aircraft Scenario 4.

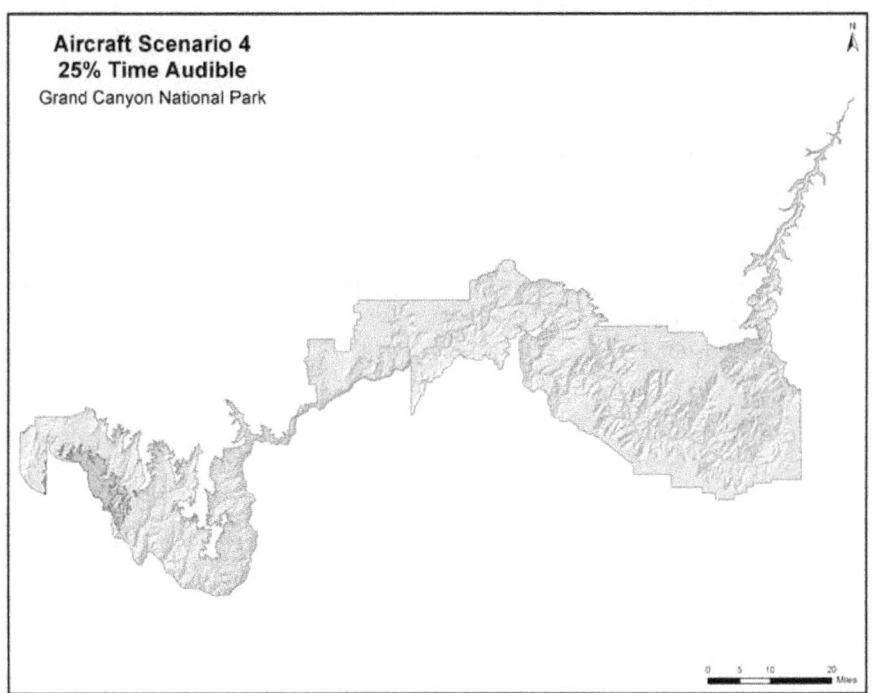

Figure 28. 25 %$T_{Aud,}$ Aircraft Scenario 4, INM
25% TAud = 1.9 % of Park

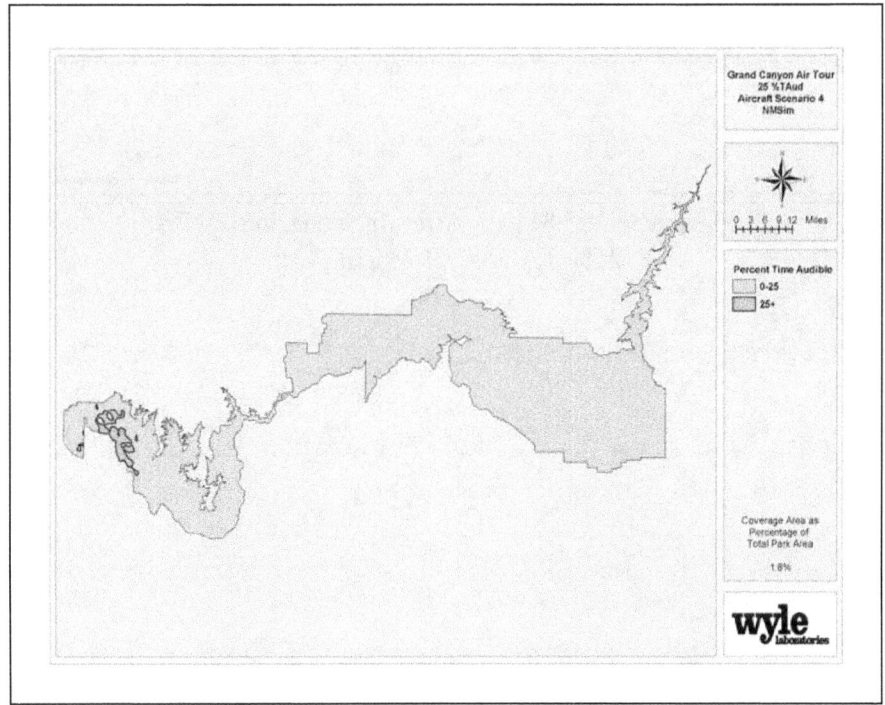

Figure 29. 25 %$T_{Aud,}$ Aircraft Scenario 4, NMSim
25% TAud = 1.8 % of Park

4.1.5 Aircraft Scenario 5

Note that both INM and NMSim predicted 0% impact for Aircraft Scenario 5. Accordingly, no figures are presented for this scenario.

4.1.6 Aircraft Scenario 6

Figures 30 and 31 depict the 25 % time audible contours associated with INM and NMSim, respectively, for Aircraft Scenario 6.

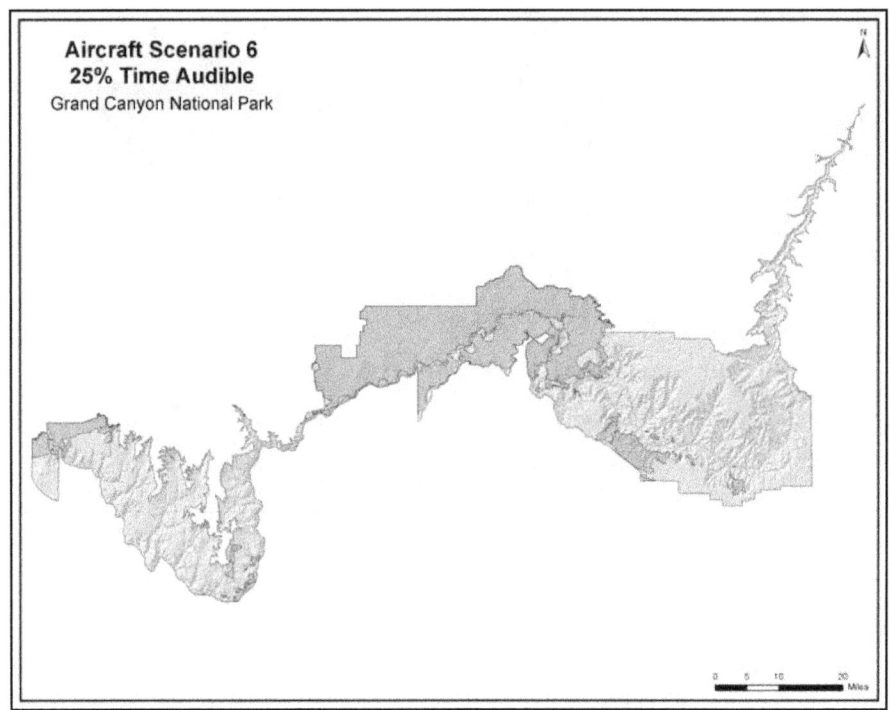

Figure 30. 25 %$T_{Aud,}$ Aircraft Scenario 6, INM
25% TAud = 29.7 % of Park

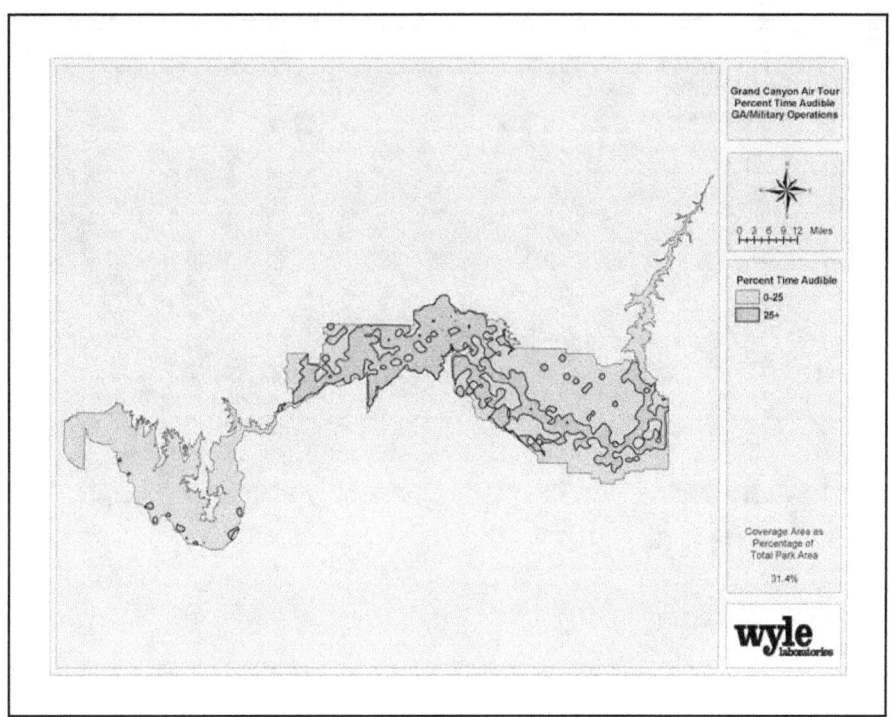

Figure 31. 25 %T_{Aud}, Aircraft Scenario 6, NMSim25% TAud = 31.4% of Park

4.1.7 Aircraft Scenarios 7 to 10

Four sets of contours are presented for the commercial, high altitude overflights. As discussed in Appendix G, there is substantial uncertainty with regard to the best approach to represent the noise from commercial high altitude overflights. Each of the four sets depict the high-altitude results of modeling the high-altitude source noise slightly differently. Figures 32 and 33 show the results computed by each model, assuming a logarithmic regression through the measured overflight data collected in support of this study (Appendix G; Aircraft Scenario 7). Figures 34 and 35 show the results computed by each model, assuming the upper 95% of the logarithmic regression through the measured data (Aircraft Scenario 8). Figures 36 and 37 show the results computed by each model assuming the lower 95% of the logarithmic regression through the measured data (Aircraft Scenario 9). Figures 38 and 39 show the results using theoretically derived NPD data for the commercial overflights (see Appendix D; Aircraft Scenario 10).

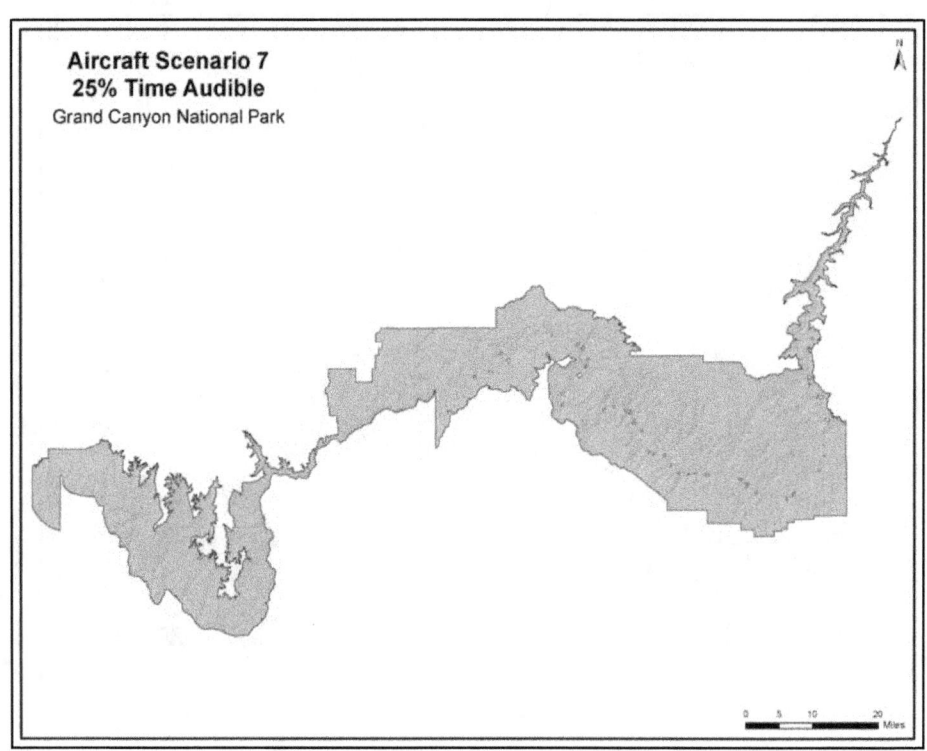

Figure 32. 25 %T_{Aud}, Aircraft Scenario 7, INM
25% TAud = 94.6 % of Park

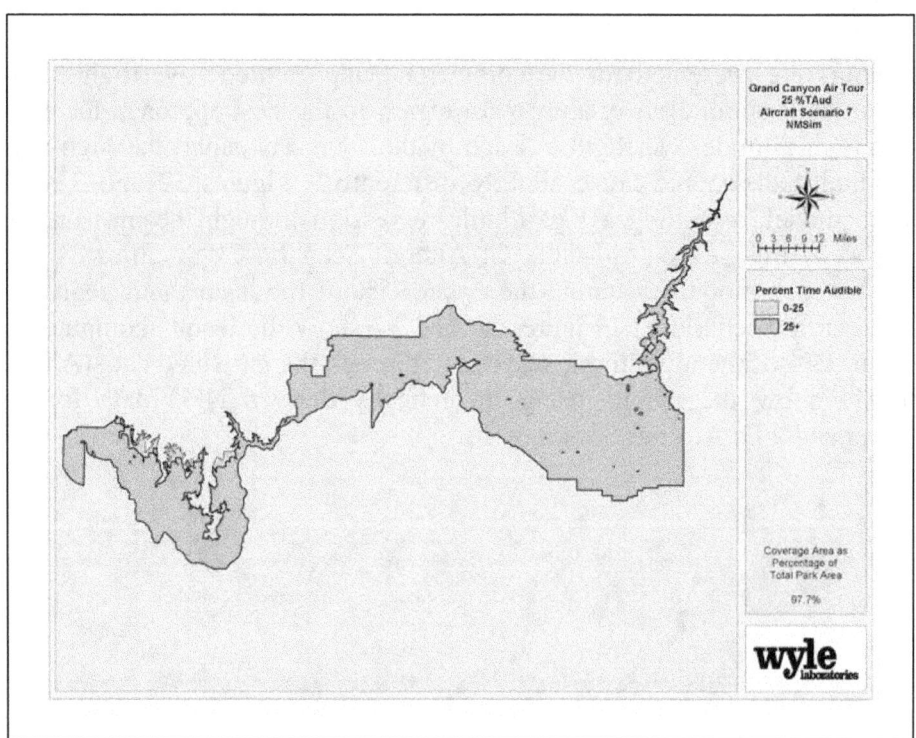

Figure 33. 25 %T_{Aud}, Aircraft Scenario 7, NMSim
25% TAud = 97.7 % of Park

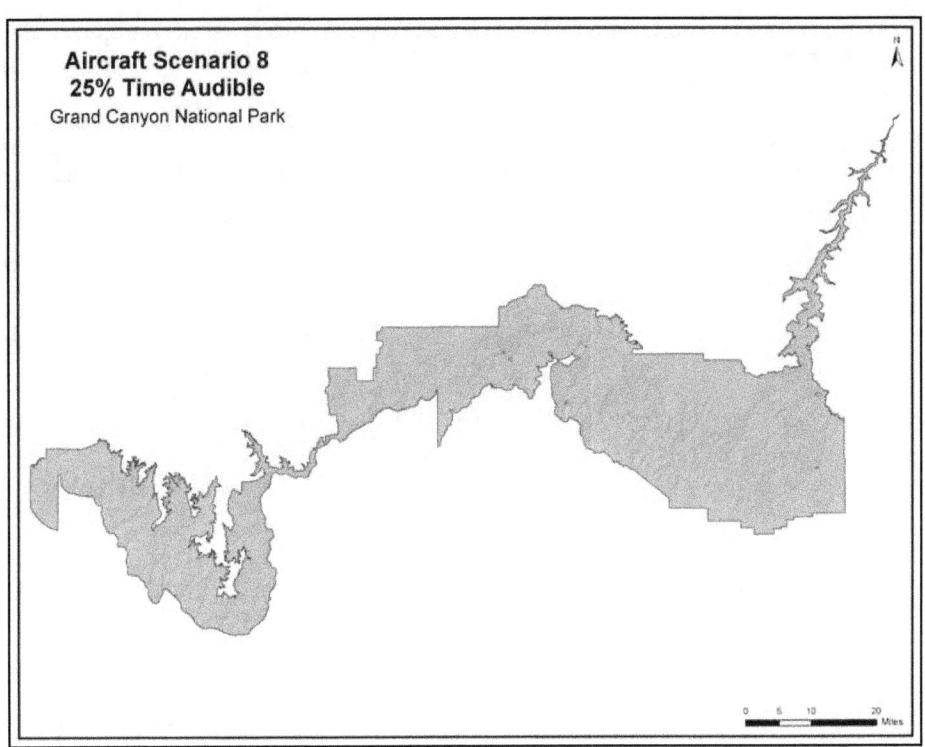

Figure 34. 25 %T$_{Aud,}$ Aircraft Scenario 8, INM
25% TAud = 99.9 % of Park

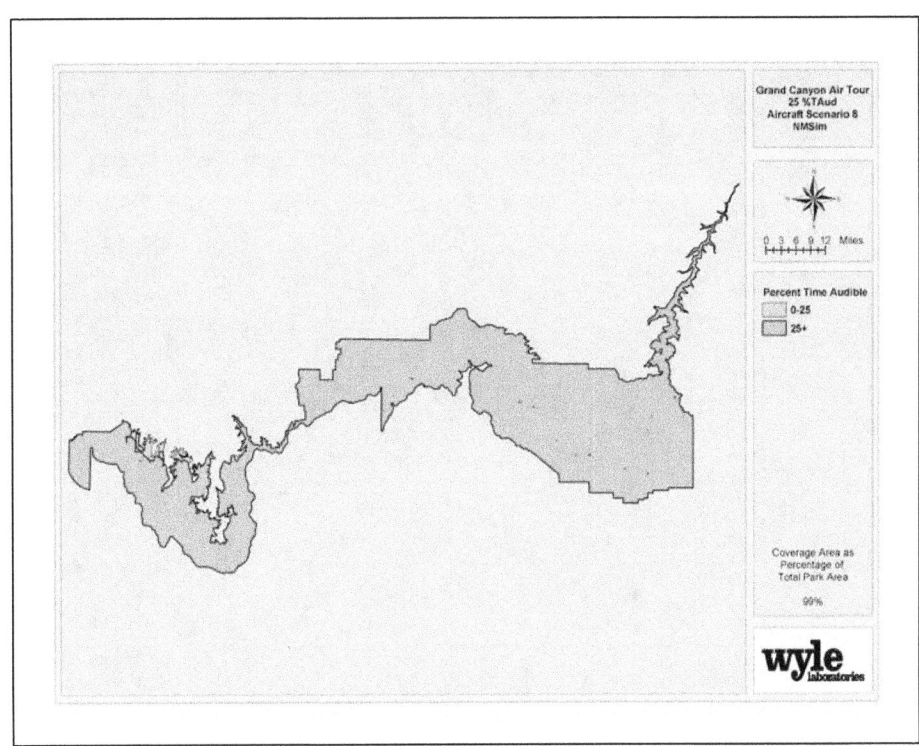

Figure 35. 25 %T$_{Aud,}$ Aircraft Scenario 8, NMSim
25% TAud = 99.0 % of Park

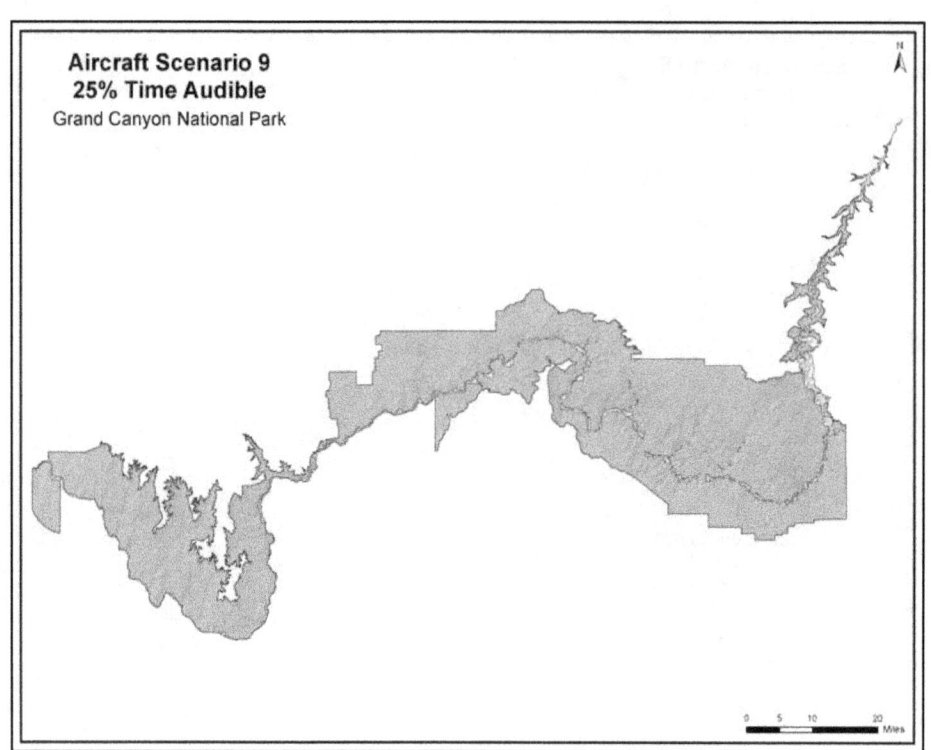

Figure 36. 25 %T$_{Aud,}$ Aircraft Scenario 9, INM
25% TAud = 95.8 % of Park

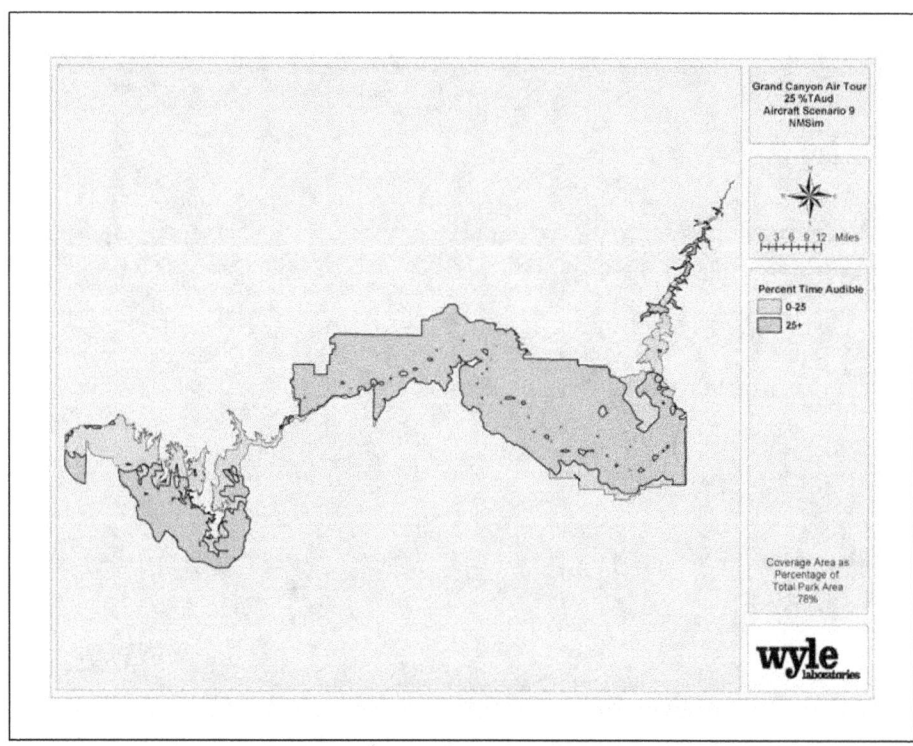

Figure 37. 25 %T$_{Aud,}$ Aircraft Scenario 9, NMSim
25% TAud = 78.0% of Park

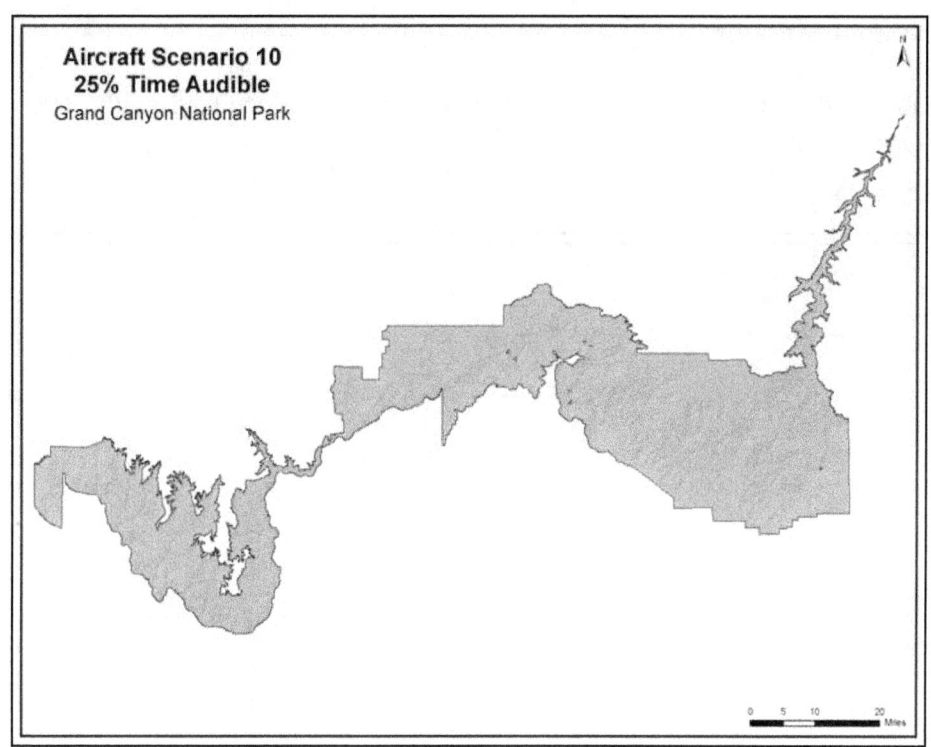

Figure 38. 25 %T_{Aud}, Aircraft Scenario 10, INM
25% TAud = 100 % of Park

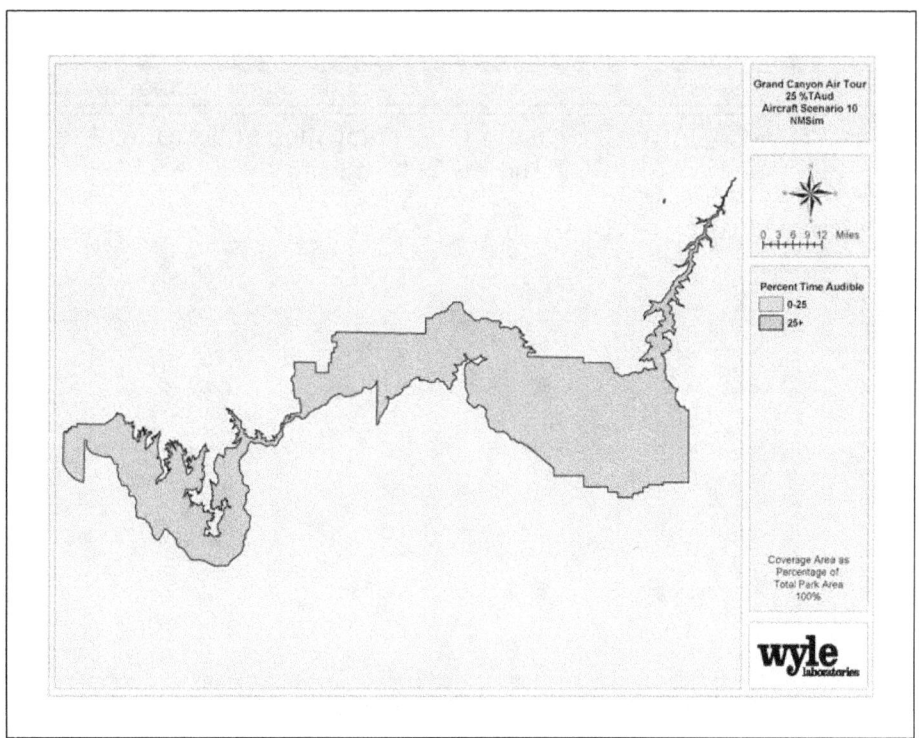

Figure 39. 25 %T_{Aud}, Aircraft Scenario 10, NMSim –
25% TAud = 100 % of Park

4.2 Aggregate Contributions

Figures 40 and 41 depict the 25 % time audible contours associated with INM and NMSim, respectively, for the combination of Scenarios 1 through 5.

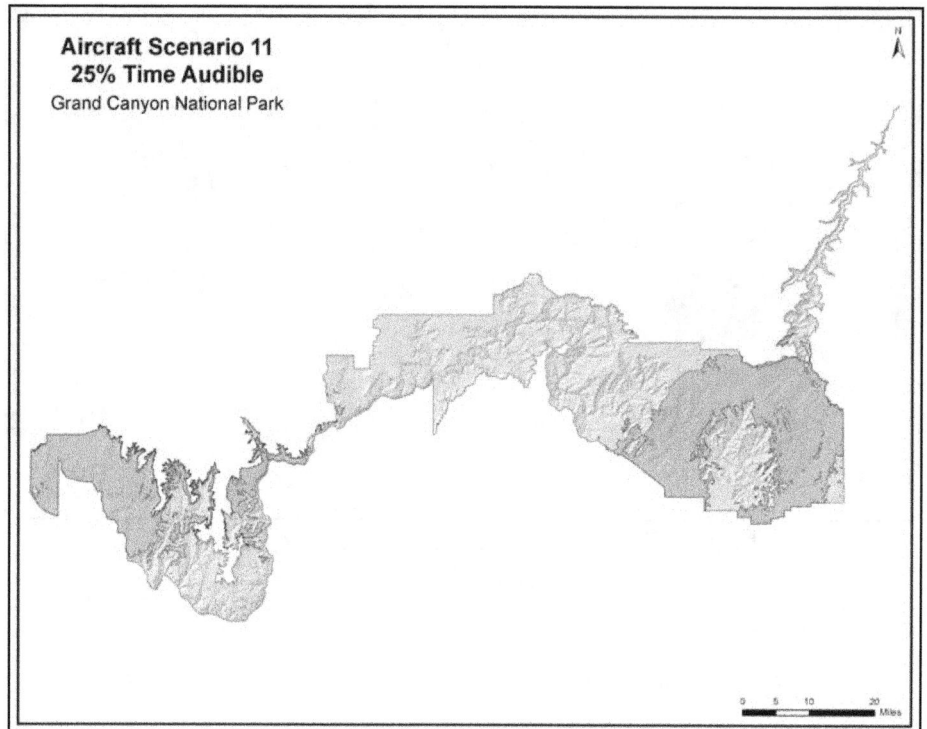

Figure 40. 25 %$T_{Aud,}$ Aircraft Scenario 11: Combination of Scenarios 1 to 5, INM 25% TAud = 37.1% of Park

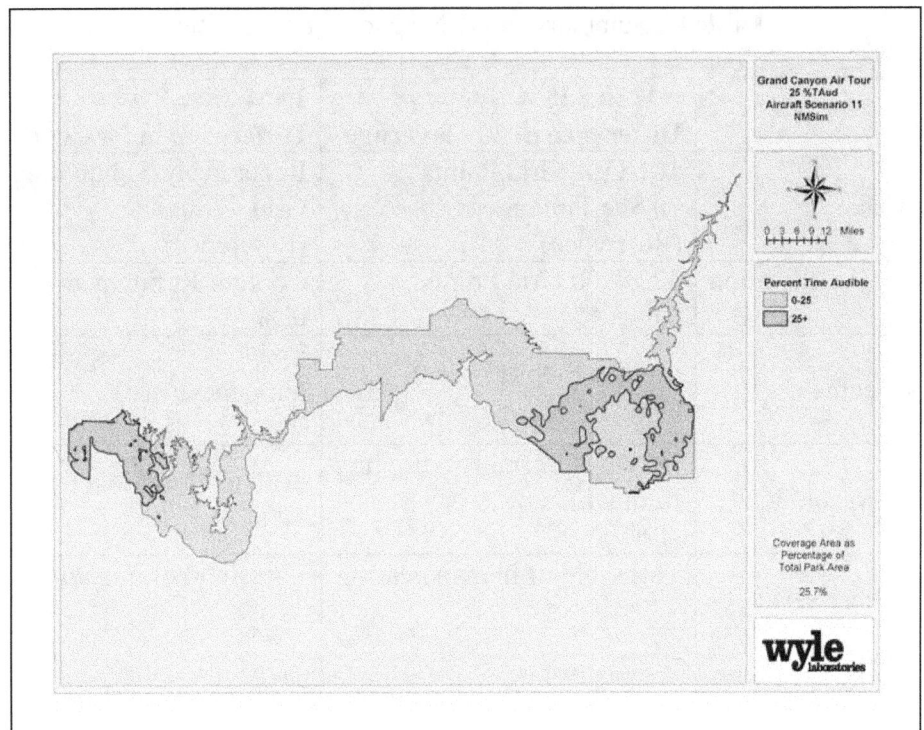

Figure 41. 25 %T_{Aud}, Aircraft Scenario 11: Combination of Scenarios 1 to 5, NMSim
25% TAud = 25.7 % of Park

4.3 Contour Sensitivities

This section examines the sensitivities of the model output to various control strategies that could possibly be introduced on the commercial air tours, and air tour related flights, including quiet technology and reduction of operations by substituting smaller aircraft with larger ones, having more seat capacity. The commercial air tour operations used in the comparative analysis described in Section 4 were based on air tour operations during the average-day of a high activity month, or specifically August 31, 2003. To understand how an increase in operations effects the audibility contours, runs were also made with both models for doubling of air tour operations. The results of these runs are summarized in Table 8.

Table 8. Summary of GCNP Model Sensitivities

	NMSim 25% TA Difference in % Coverage	INM 25% TA Difference in % Coverage
Base Case	161X100 Grid Points 10 Sec Time Steps (Reference)	Recursively-Subdivided Grid (Reference)
Contour Generation Studies	322X200 Grid Points 0%	Contour Refinement 0 %
Track Segments	½ Time Steps (5 secs) 0%	Track Segmentation (simulate 5 sec) 0%
Terrain Resolution	Halving Terrain Sampling from DLGs -1.6%	3CD to Grid Float -4%
Ambient Effects	Threshold of human hearing +29%	Threshold of human hearing +13%
Quiet Technology – Aircraft Replacement (1-for-1 aircraft)	-14.0%	-9%
Quiet Technology – Aircraft Replacement (seat-swap)	-19.9%	-11%

4.4 Margin of Safety (Contours with uncertainties)

The NPS has defined *restoration of natural quiet in GCNP* to occur when aircraft are audible in less than 50% of the park 25 percent of the time in which it is open (7 am to 7 pm). Though INM and NMSim predict similar time audible levels in GCNP, the uncertainty around those contour predictions has not been quantified. The uncertainty with such predictions is an important parameter for both NPS and FAA to understand, particularly in cases where the models indicate values close to the NPS goal of restoration. The purpose of the margin of safety assessment is to provide a first-order approximation of the lower bound to uncertainty around the GCNP contours generated in support of this study. The uncertainty assessment is included in this study primarily for the benefit of the FAA and NPS as part of the ADR process.

In assessing contour uncertainty, the authors determined that it was not practical to conduct a comprehensive statistical analysis of all the models' uncertainties in support of this study. Consequently, it was decided to base the uncertainty limits on common rules of thumb. Figure 41 presents a comparison of measured time audible values simultaneously logged at the same site by two separate observers in the GC MVS. The average agreement between the two observers is quite good, however, the scatter about the average is substantial. The scatter provides a quantification of uncertainty associated with measuring aircraft audibility in GCNP. This uncertainty may in fact be representative of measurements in other parks as well.

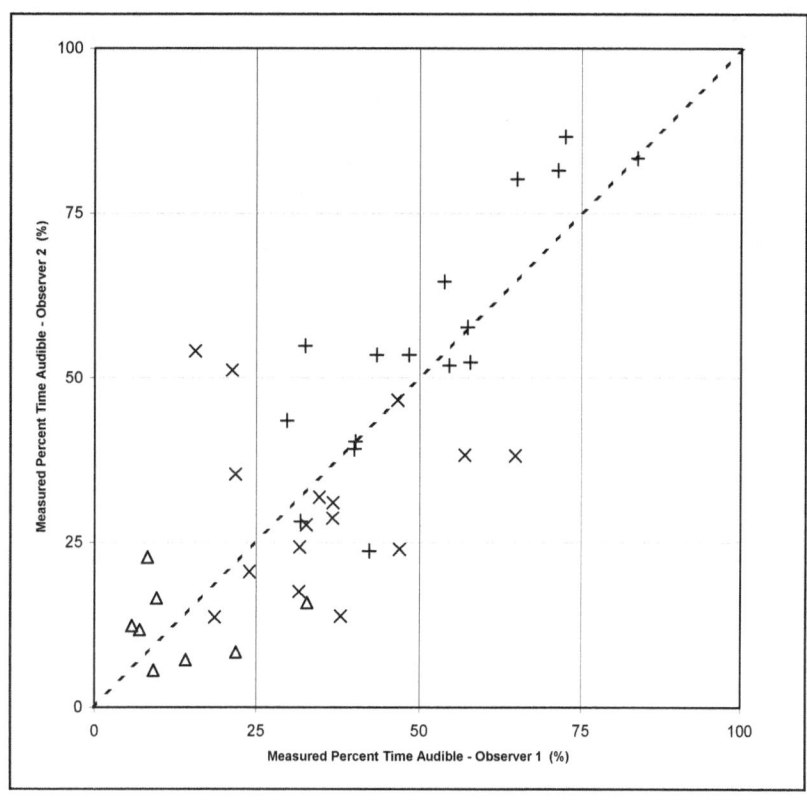

Figure 42. Comparison of Measured Audibility in GCNP

It is unreasonable to expect any model to provide time audible predictions with less uncertainty than that represented by the scatter in Figure 42. Consequently, the data presented in the figure were used to compute an upper and lower bound to the 95% confidence interval. Specifically, the GCNP MVS report attributed approximately 4% error to the measurement errors of the type highlighted in Figure 42. Therefore, to provide a measure of the uncertainty in the models' predictions of the percentage of GCNP area restored to natural quiet, the areas associated with the 29% (25% + 4%) and 21% (25% - 4%) time audible contours were computed. The results are summarized in Table 9 for aircraft Scenario 11. Similar error bounds are expected with NMSim.

Table 9. Summary of Lower Bounds to Contour Area Uncertainty
GCNP Aircraft Scenario 11

Time Audible Contour	Change in Percentage of GCNP Restored (%)
	INM
27% (+95% CI)	+2.6
25% (GCNP Baseline)	(reference)
22% (-95% CI)	-2.3

5. Model Usability

The purpose of this section is to address the usability of the two models. Noise modeling in the GCNP is a unique situation in that the modeling has traditionally been performed by two organizations, FAA and the NPS, with assistance from Volpe and, most recently, Wyle. However, Congress has mandated that ATMPs be developed in National Parks with air tours, currently estimated at over 100. Although it is possible for Volpe and Wyle to conduct noise modeling internally for all of these parks, it may not be desirable. Consequently, if it is desired that multiple organizations conduct ATMP analysis, the model to be used must: (1) be publicly available; (2) be user friendly; (3) generate credible results with reasonable runtimes; and (4) be relatively turn-key – it cannot require a substantial amount of custom programming and hand holding to transform preliminary output into a form that is needed to support ATMPs. In short, the model must be useable by the general aviation noise modeling community. Although the judgment of usability is somewhat qualitative, this section attempts to highlight some of the key issues to consider in understanding usability.

5.1 Data Requirements

Input data requirements for the two models are similar. Specifically, the models both require aircraft operations data: trajectories, numbers, and types. As discussed in Section 2.1.2, most of the aircraft flying in the National Parks are represented in the database of the INM, and for the few that are not, there are FAA-approved substitutions. The NMSim database of parks-specific aircraft currently includes all aircraft operating at Grand Canyon, with the exception of the MD900. Some of this noise data were derived from the INM database. Future park modeling using NMSim is likely to require import/derivation of additional data. The process of importing/deriving INM aircraft for NMSim involves the calibration of resultant NMSim sound level data to the INM NPD level at a reference distance. Effort to import and calibrate INM data is modest (about 15 minutes per aircraft) but has thus far involved the use of proprietary tools (see Section 2.1.1 for limitations on the derivation process). INM also supports a user-defined aircraft definition function, which allows new data not already a part of the standard database to be easily imported by the user.

INM is consistent with the internationally-accepted and peer-reviewed guidance of the Society of Automotive Engineer's (SAE) SAE-AIR-1845 "Procedure for the Calculation of Airplane Noise in the Vicinity of Airports" and contains its aircraft performance equations for fixed wing aircraft. NMSim does not include an aircraft performance model. Neither model contains performance equations for rotorcraft. Both models can use performance data directly supplied by tour operators in the National Parks. In lieu of direct performance data, INM can generate fixed wing power profiles directly from the SAE-AIR-1845 equations. NMSim can import power profiles (as well as flight tracks) generated by INM.

Another key user input data requirement is ambient noise, or underlying ambient maps. Data to support these maps are currently being collected jointly by the FAA and NPS, with the assistance of Volpe, and both models are capable of utilizing these data directly. The data are required to model time audible, time above or change in exposure metrics. The INM implements multiple ambient sound level data values for an analysis via a check-box in the graphical user interface (GUI). When this option is selected, INM utilizes two ASCII text files – an ESRI ASCII grid file containing A-weighted levels, and another file that assigns spectra to the values in the grid file. NMSim implements multiple ambient sound level data via an ESRI ASCII grid file containing unique indices, and a file defining ambient spectra for each index. The ambient case to be used by NMSim is selected from a pull-down menu.

5.2 User Interface

Both INM and NMSim are menu-driven, Windows programs. The graphical user interface (GUI) of INM has had a similar look and feel since the release of Version 5.0 in the mid 1990s (see timeline in Figure 1 of Section 1.1.1). Consequently, much of the noise modeling community is familiar with its input/output. Although NMSim has yet to be released, other than as a beta version, the GUI is relatively intuitive and model usage is fairly easy to pick up after minimal exercising.

5.3 Outputs

INM contour and grid point outputs are generated in a tightly integrated, closed system. In the case of contours, an underlying grid file of receptors is generated by the program, the software internally shells out to the NMPlot contouring routine, and the resultant contour is computed and rendered within INM. The receptor file used for contouring within INM consists of a base grid of receivers of user-defined density, which is automatically subdivided, based on a user-defined precision. INM grid files can be exported to DXF, Shape Files, MapInfo Files, and NMGF formats.

NMSim generates a regularly-spaced grid file, at a resolution specified by the user. This can be either an ESRI ASCII grid, readable by GIS software, or an NMGF file that can be displayed by NMPlot. From there the contours can be exported to DXF or Shape Files.

The advantage of bringing a grid file directly into NMPlot is that it allows the user the ability to take advantage of all of the functionality within NMPlot. The downside to this approach is that NMPlot is somewhat complex for the casual user. The program allows the user to tweak many parameters associated with the contouring algorithm, which can result in seemingly different contour results generated from the same grid file. This downside is the primary reason that INM integrates NMPlot directly within its GUI, although INM like NMSim provides output files to support standalone use of NMPlot.

Both INM and NMSim generate grid files that can be imported into standard GIS systems for contouring and other related analyses. GIS systems can be used to compute contour areas, as well as contour areas within a designated boundary, such as a National Park boundary. In addition to supporting this capability externally using the grid file, INM provides contour area data directly in the output data. NMSim further has the capability of generating color renderings of any of the grid-oriented metrics, and animations of the acoustic time history.

As discussed in Section 2.3, both models can compute a wide range of noise and time-related metrics. It is expected, based on previous studies in support of the National Parks, that the following metrics may be required: day-night average sound level, time audible, time above a user-defined threshold, change in exposure, maximum A-weighted sound level, and equivalent sound level over a user-defined period. INM and NMSim compute all of these metrics.

5.4 Implementation

This section discusses issues of modeling in both INM and NMSim that are not discussed in the documentation of the models. The primary focus is on undocumented utility programs required by the users of each model.

INM: The INM has two guides to assist the user. The INM Users Guide discusses the methods of entering data in the INM. The INM Technical Manual discusses the underlying algorithms of the INM. These documents discuss standard use of the INM for aircraft operations around airports; they do not discuss issues specific to modeling operations around National Parks, where the primary concern is overflight noise, not departure and arrival noise. However, substantial guidance on the use of INM for modeling in the National Parks is included with the Version 6.2 release notes (see Appendix B).

Input track information is typically available in latitude and longitude coordinates. This type of information cannot be directly used in the INM; the data must be converted to a local Cartesian coordinate system that has its origin at a user defined point in the study. An example is the FAA-provided latitude and longitude flight track information for the Grand Canyon air tours. These coordinates are found on the VFR map used by the air tour pilots. For the current study these coordinates are converted into the local coordinate system using a utility program. In addition to importing the points that determine the track, a second utility program was used to insert intermediate points between existing points so that the long flight track segments could be broken down into shorter segments. For the final INM runs used in this study, the final flight track segment size was reduced to 2000 foot spacing. This spacing corresponds to a time segment of 10 seconds for an aircraft traveling 118 knots, approximating the time step utilized in NMSim for this study. Both the function of converting the coordinates and sub-segmenting the flight tracks, which were performed using utility software external to the INM, can be accomplished directly within the INM GUI, although as with many GUI features for both of the models, using the GUI is a bit more tedious when dealing with large amounts of data.

Users of the INM can import one of three terrain data types: the 3CD files available to previous versions of the INM and discussed in the Users Guide, GridFloat data for various refinement, and DEM data. To make use of these data, the user selects which of these types to use via the GUI and enters the location of the files in the directory structure of the computer. The INM determines which of the terrain files to open to encompass the study area. These higher resolution GridFloat and DEM data are available free of charge from the U.S. Geological Survey (USGS) National Map Seamless Data Distribution System (http://seamless.usgs.gov).

INM requires no utility software to analyze the generated contours. It develops contours via its internal link to NMPlot; additionally, standard GIS-ready output data such as DXF CAD and ShapeFiles may be exported.

NMSim: Like the INM, NMSim has a Graphical User Interface (GUI) that helps to facilitate data entry. Also like the INM, this interface can become tedious for data entry, and so users can by-pass the GUI and enter data directly in files used by NMSim.

NMSim track information can be entered in either geographical coordinates or in the coordinates used in the study. The primary input is via the Track Builder. This utility program allows the user to enter data graphically, or by typing specific values into dialog boxes. The trajectory file ("TRJ") is stored in tabular ASCII format. Users can create TRJ files outside of the GUI. NMSim tracks can be generated for ground vehicles as well as for aircraft.

INM and NMSim use different methods of assigning aircraft performance information to individual flight tracks. In INM, performance information is defined as speed-altitude-power profiles along straight paths, and flight tracks are defined as 2-D paths along the ground. Different aircraft with different performance profiles can use the same tracks. INM merges these internally into trajectories that contain full 3-D path and performance data. NMSim, like its predecessor NoiseMap, requires that users directly provide merged trajectories with both track and profile data. In practice, this means that the INM's data entry is more efficient than NMSim's when multiple aircraft types operate on a single track (e.g. the air tour operators in GCNP that operate different aircraft types on the same, known tracks), but user input of data into NMSim becomes more efficient when each individual track has only a single operation (e.g., the civil over-flights tracks taken from radar data, where each radar track has only a single associated operation).

Because of the inherent tediousness of using any GUI to enter copious amounts data, the developers of NMSim have created proprietary tools to convert data already entered in the INM to a format that can be converted to the underlying NMSim ASCII files. This method uses the INM to combine the track and performance information into the unified format needed by NMSim. While the INM is not required to do this unification, using the INM as a pre-processor for NMSim has proven expedient for this project. Importing profiles and tracks already processed into INM form by Volpe ensured that both models were using identical inputs.

The terrain entry for NMSim is similar in concept to that of INM. The user imports topography files that cover the area of interest, then defines the study area by its geographic corners. The imported data are visible as the user selects the corners. The user can overlay background layers from DLG or ESRI shape files during this process, as well as later in the analysis.

Elevation data file formats currently handled by NMSim include USGS DEM, USGS DLG and DTED. All resolutions of each type are supported, but it is strongly recommended that only one type and one resolution be used in a given study - the data sources are not entirely consistent. Elevation data in ESRI ASCII grid format may also be used. The user specifies the resolution desired for the elevation file, which is derived from the raw terrain data. The elevation file is stored as type "ELV," and is in standard NMGF V1.0 format. If there is a need, like in INM, users can directly write terrain data files from unique data sources.

As noted above, GUI data entry can become tedious for large cases. The presence of Windows API calls in an executable can also slow processing, even for parts of the code not directly using them. For large cases, NMSim has a batch mode. This batch file system can control the entire NMSim process so that the user may by-pass the GUI entirely, if desired, or use it for efficient calculations after setting up a case in the GUI.

For studies at GCNP, the NMSim batch file system calculates the impact of each individual track operation using one set of batch files; each batch file corresponds to one unique track-operation. A master batch file calls each of these subordinate batch files; this master batch file determines the total number of operations that are modeled. This master batch file corresponds to a "case" in an INM study. Because of the size and number of the resultant spectral time history grid ("TIG") files, which can easily grow to several GB each, the individual spectral time history grid files are immediately used to calculate d-prime time histories, and may then be discarded to save storage space.

The results of NMSim include time-history ASCII text files that can be graphically viewed in the NMSim Visualizer. The Visualizer can export these images to standard graphic file formats, and can generate animations of the acoustic time history. These ASCII files can also be exported to GIS post-processing systems. Because they are ASCII, no special decoding is needed to read them for other purposes.

NMGF grid files generated by NMSim are internally forwarded to NMPlot, for analysis and post-processing similar to INM's incorporation of NMPlot.

5.5 Other

INM, currently in its 26th year of public distribution, has been disseminated to over 800 individuals and organizations in more than 50 countries world-wide. The user base includes government, academia, aviation industry and commercial engineering. While INM is often used in support of required state and federal environmental analyses, it is also used widely for research. A standardized Research Version is exercised regularly in support of technical working groups and government agencies. The INM is distributed with both a User's Guide and a Technical Manual for all major releases (i.e., 5.0, 6.0, etc.). Interim releases are disseminated with technical release notes which identify any changes with respect to the current documentation. INM training courses are provided by private industry in both the United States and Europe. Free technical support is provided by the development team to all users of the model.

NMSim originated in the 1990s as a R&D tool for NoiseMap development, in particular to support the addition of topographic effects to NoiseMap 7. Because of this origin, it is structured to use modern propagation algorithms. Algorithm development and validation was conducted under the guidance of an international NATO-CCMS committee. In 2001, NPS contracted with Wyle Laboratories to develop NMSim into a user-friendly GUI version. Development was supported by an international beta test team that included industry, consultants and government agencies. When released, NMSim source code will be available under the GNU General Public License. The executables will be accompanied by a User's Manual and on-line help. Wyle will maintain a NMSim support forum site. During beta testing, several training sessions were conducted. It is expected that Wyle will offer NMSim training courses.

NMSim was scheduled for public release in early 2004. That release has been deferred until the completion of the effort presented in this report, in the expectation that lessons learned would be incorporated into the release. Both NMSim and INM 6.2 beta have benefited from updates developed as analysis under the current effort proceeded.

Presenting a meaningful comparison of runtime for the two models presents a challenge since the process of developing a contour within each system is quite different, as discussed herein. In an effort to try and ensure the most meaningful comparison of runtime for the core acoustics in the two model, audibility values were computed for a regularly-spaced grid of 32-by-32 receivers spanning the entire GCNP analysis window. The commercial air tour case was run for the two models and a comparison of core computational runtime is presented in Table 10. As can be seen, core computational runtime for INM and NMSim is comparable.

Table 10. Comparison of Core Computational Run Time (minutes)
32-by-32 Grid of Receivers

INM	NMSim
26	20

6. Summary of Findings and Recommended Improvements to the Models

Following is a brief, bulletized summary of the findings of the current study:

Defensibility:
- The components of both INM Version 6.2 and NMSim are based on well-established physics, and have been field validated. (See Section 2)
- INM is based on internationally-accepted, peer-reviewed SAE standards; NMSim is based on peer-reviewed international journal articles and related research reports. (See Section 1.1)
- INM has withstood numerous legal challenges since its inception.

Flexibility:
- INM includes a comprehensive aircraft performance model, which supports terminal area performance modeling, as well as low-altitude performance modeling (the case in the National Parks). NMSim does not model aircraft performance. It contains GUI tools for entering flight paths, but it is up to the user to define performance. (See Section 2.5)
- Given that NMSim will be distributed with open source code, it may offer some flexibility over INM for the more knowledgeable users who have the capability to tailor the source code to support their analyses. Terms of NMSim's open-source license require that users disclose any changes made to the source code, which should allow for careful audits of code changes (see Section 5).
- NMSim includes the Scheduler, which allows the user to account for overlapping noise events. Similarly, INM has a utility program which empirically adjusts modeled audibility to take into account overlapping events. The INM empirical adjustment is based on NPS acoustic measurements at GCNP, and its applicability outside of the GCNP environment remains unclear. It is not, however, expected that this will be an important issue in most parks, since the number of air tour flights are much fewer than for GCNP. NMSim scheduler results and the INM empirical model agree well with each other for the GCNP MVS. (See Section 2.5)

Usability:
- Both models include comprehensive user guidance material, including User's Guide and Technical Manual (not yet available for NMSim), which chronologically document all enhancements to the model. (See Section 5)
- INM provides technical support to all registered users. (See Section 5)
- INM has courses available for interested users. Wyle plans to offer training courses for NMSim. (See Section 5)
- Core computational runtime for INM and NMSim is comparable. (See Section 5)

Maintenance and Development:
- Both models have annual budget streams and are continually updated and improved.

Noise Database:

- The database in INM Version 6.2 includes noise and performance information for the vast majority of tour aircraft flying in the National Parks, based on a recently conducted survey; as it currently stands the database in NMSim includes all but one of the tour aircraft that are operating in GCNP. NMSim-compatible data can be derived from INM NPD data. There are, however, elements of the derivation process that differ from INM database development process and result in differences when compared with the original INM NPD data, although analyses conducted since the October version of this document have led to NPDs that generally compare well. NMSim sources derived from INM NPDs are considered low resolution because the available NPD data have been integrated, and only SEL, L_{ASmx}, EPNL, L_{PNTSmx} and spectrum at time of L_{ASmx} data are available (see Section 2.1).

Model-to-Model Comparisons:
- The noise databases for the six aircraft used by the two models compare reasonably well. SEL values are within 2 dB out to a distance of 25,000 ft, and maximum sound levels are within 3 dB out to the same distance, with the exception of the DHC-6, which was different by almost 6 dB at 25,000 ft.
- The contour areas for GCNP agree reasonably well, with INM Version 6.2 generally computing slightly higher levels of audibility compared with NMSim.
- Substantial gains have been made with regard to understanding model-to-model differences; and many of those differences have been reduced or eliminated. However, when comparing INM Version 6.2 and NMSim, there still remain some differences, particularly with point-to-point comparisons.

Accuracy:
- Both INM Version 6.2 and NMSim are performing equally well, on average, when compared with the "gold standard" audibility data measured in the GCNP MVS. (See Section 3.3)

The authors of this document did not think it would be appropriate to make recommendations with regard to specific model usage, and as such have attempted to be non-judgmental herein. However, there are several recommendations that came out of this study related to model improvements, which should be noted.

Recommendations:
- The INM database should be expanded to include the Beech C99, Cessna 182R, the Cessna 208, the Robinson R-44 helicopter and possibly the Fokker F-27. These aircraft are fairly prominent in regard to park air tour flights, and are not well represented in the INM.
- The database of NMSim should be expanded to include many of the aircraft that are currently flying in the National Parks. Ideally, this would be done through a commensurate measurement study, using a tower-based microphone system to properly allow for hemisphere development, but a possible, less preferred alternative would be to convert existing INM data.
- The INM needs to adopt an algorithm for taking into account the effects of acoustically hard ground, once this has been adopted by SAE; both models need an algorithm for accounting for propagation over complex, mixed ground.
- Inclusion of source directivity in INM should be investigated. Source directivity was not significant for Grand Canyon, where terrain shielding dominated propagation from distant portions of flight tracks, but may be important in parks with less rugged terrain.

7. References

[1] Miller, N.P., et. al., Aircraft Noise Model Validation Study, HMMH Report No. 295860.29, Harris Miller Miller and Hanson, Burlington, MA, January 2003.

[2] AIR 1845, Procedure for the Calculation of Airplane Noise in the Vicinity of Airports, March 1986.

[3] AIR 1751, Prediction Method for Lateral Attenuation of Airplane Noise during Takeoff and Landing, March 1981.

[4] AIR 866A, Standard Values of Atmospheric Absorption as a Function of Temperature and Humidity, Prepared by the SAE Committee A-21, Aircraft Noise Measurement, Report No. ARP 866A, Issued 8-31-64, Revised 3-15-75.

[5] ICAO, "Recommended Method for Computing Noise Contours around Airports", Circular 205-AN/1/25, March 1987.

[6] Methodology for Computing Noise Contours around Civil Airports, Volume 2: Technical Guide, Draft Version 6.0, European Civil Aviation Conference, ECAC. CEAC Doc 29R, Proposal by AIRMOD Technical Subgroup, May 12, 2004.

[7] Fleming, et. al., Integrated Noise Model (INM) Version 6.0 Technical Manual, FAA Report Number FAA-AEE-02-01, Federal Aviation Administration: Washington, DC, January 2002.

[8] Reherman, Clay; Roof, Christopher; Fleming, Gregg; Senzig, David; Read, David; and Lee, Cynthia, Fitchburg Municipal Airport Noise Measurement Study: Summary of Measurements, Data and Analysis, Report No. DTS-34-FP301-LR3/DTS-34-FA365-LR1, U.S. DOT, Research and Special Programs Administration, John A. Volpe National Transportation Systems Center Acoustics Facility: Cambridge, MA, September 2003.

[9] Reherman, Clay; Roof, Christopher; Fleming, Gregg; and Read, David, Crows Landing Noise Measurement Study: Summary of Measurements, Data and Analysis for the MD 600N Helicopter, Report No. DOT-VNTSC-FAA-03-11, U.S. DOT, Research and Special Programs Administration, John A. Volpe National Transportation Systems Center Acoustics Facility: Cambridge, MA, May 2004.

[10] Technical Memo, NPD Data for MD900 and DHC-6 in Support of INM and GCNP Modeling Effort, Fleming, Gregg, March 19, 1999.

[11] Senzig, D.A., Memorandum to Mr. Ted. Farwell, Cessna Aircraft Company, Cessna 172R data for the Integrated Noise Model, August, 29, 2000.

[12] Reherman, C.N., Roof, C.J., Fleming, G.G., Read, D.R., Crows Landing Noise Measurement Study: Summary of Measurements, Data and Analysis for the MD 600N Helicopter, DOT-VNTSC-FAA-03-11, May 2004.

[13] Rickley, E.J., et. al., Aerospatiale Astar SA350D, FAA Report Number FAA-EE-84-05, Federal Aviation Administration: Washington, DC, 1984.

[14] Rickley, E.J., et. al., Aerospatiale Dauphin SA365N, FAA Report Number FAA-EE-84-02, Federal Aviation Administration: Washington, DC, 1984.

[15] Rickley, E.J., et. al., Aerospatiale Gazelle SA341G, FAA Report Number FAA-EE-79-03, Federal Aviation Administration: Washington, DC, 1979.

[16] Rickley, E.J., et. al., Aerospatiale Puma SA330J, FAA Report Number FAA-EE-79-03, Federal Aviation Administration: Washington, DC, 1979.

[17] Rickley, E.J., et. al., Aerospatiale Twinstar SA-355F, FAA Report Number FAA-EE-84-04, Federal Aviation Administration: Washington, DC, 1984.

[18] Rickley, E.J., et. al., Augusta A-109, FAA Report Number FAA-EE-81-16, Federal Aviation Administration: Washington, DC, 1981.

[19] Rickley, E.J., et. al., Bell 212, FAA Report Number FAA-EE-79-03, Federal Aviation Administration: Washington, DC, 1979.

[20] Rickley, E.J., et. al., Bell 222, FAA Report Number FAA-EE-84-01, Federal Aviation Administration: Washington, DC, 1984.

[21] Rickley, E.J., et. al., Boeing/Vertol CH-47D Chinook, FAA Report Number FAA-EE-84-07, Federal Aviation Administration: Washington, DC, 1984.

[22] Rickley, E.J., et. al., Boelkow BO-105, FAA Report Number FAA-EE-79-03, Federal Aviation Administration: Washington, DC, 1979.

[23] Rickley, E.J., et. al., Hughes 500 D/E, FAA Report Number FAA-EE-84-03, Federal Aviation Administration: Washington, DC, 1979.

[24] Rickley, E.J., et. al., Sikorsky S-61, FAA Report Number FAA-EE-79-03, Federal Aviation Administration: Washington, DC, 1981.

[25] Rickley, E.J., et. al., Sikorsky S-65, FAA Report Number FAA-EE-79-03, Federal Aviation Administration: Washington, DC, 1984.

[26] Rickley, E.J., et. al., Sikorsky Blackhawk UH-60A, FAA Report Number FAA-EE-81-16, Federal Aviation Administration: Washington, DC, 1981.

[27] Rickley, E.J., et. al., Sikorsky S-76 Spirit, FAA Report Number FAA-EE-84-06, Federal Aviation Administration: Washington, DC, 1984.

[28] Olmstead, et. al., Integrated Noise Model (INM) Version 6.0 User's Guide, FAA Report Number FAA-AEE-99-03, Federal Aviation Administration: Washington, DC, September 1999.

[29] Ikelheimer, B., and Plotkin, K.J., "Noise Model Simulation (NMSim) User's Manual," Wyle Report WR 03-09, October 2004.

[30] Plotkin, K.J., "The Role of Aircraft Noise Simulation Models, " Inter-Noise 2001 Proceedings, August 2001.

[31] "Effects of Topography on Propagation of Noise in the Vicinity of Airfields," NATO CCMS Report Number 200, 1994.

[32] 'Aircraft Noise Propagation Over Varying Topography: Measurements Made at Narvik Airport - Framnes Norway," NATO CCMS Report, 2001.

[33] Rasmussen, K.B., "The Effect of Terrain profile on Sound Propagation Outdoors," Danish Acoustical Institute Report 111, January 1984.

[34] Rasmussen, K.B., "Outdoor Sound Propagation Near Ground Surfaces," The Acoustics Laboratory, Technical University of Denmark Report no. 45, 1990.

[35] Plovsing, B, "Aircraft Sound Propagation over Non-Flat Terrain. Development of Prediction Algorithms," AV 1015/93, DELTA Acoustics & Vibration, 1994.

[36] Plovsing, B., "Aircraft Sound Propagation over Non-Flat Terrain. Prediction Algorithms," AV 7/94, DELTA Acoustics & Vibration, 1995.

[37] Plotkin, K.J., Lee, R.A., and Downing, J.M., "An Empirical Model for the In-Flight Noise Directivity of an F-16C," Inter-Noise 96 Proceedings, 1996.

[38] Lee, R.A., Liasjo, K., Buetikofer, R., Svane, C., and Plotkin, K.J., "Noise Measurements - Modeling for Topography," Noise-Con 96 Proceedings, 1996

[39] Czech, J. J., and Plotkin, K. J., "NMAP 7.0 User's manual", Wyle Research Report WR 98-13, November 1998.

[40] Plotkin, K.J., "C-130 Noise Simulation Model," prepared for AFRL, December 1999.

[41] Plotkin, K.J., Czech, J.J, and Page, J.A., "The Effect of Terrain on the Propagation of Sound Near Airports," Wyle Research Report WR 96-20, January 1997.

[42] Plotkin, K.J., "Analysis of Acoustic Modeling and Sound Propagation in Aircraft Noise Prediction," Wyle Report WR 01-24, November 2001.

[43] Plotkin, K.J., "The Effect of Atmospheric Gradients on Aircraft Noise Contours," Paper 3aNSa2, 141st Meeting of the Acoustical Society of America, June, 2001.

[44] Plotkin, K.J., Ikelheimer, B., and Huber, J., " The effects of Atmospheric Gradients on Airport Noise Contours,", Wyle Report WR 02-26, December 2002.

[45] Page, J., Plotkin, K., and Downing, M., "Rotorcraft Noise Model (RNM 3.0) Technical Reference and User's Manual, Wyle Research Report WR 02-05, March 2002.

[46] Page, J.A., and Plotkin, K.J., "Acoustic Repropagation Technique Version 2 (ART2)," Wyle Research Report WR 01-04, January 2001.

[47] Fleming, G.G., Senzig, D.A., Clarke, J.B., "Lateral attenuation of aircraft sound levels over an acoustically hard water surface: Logan airport study," January/February, 2002. (Vol. 50, No. 1, pp 19-29, 2002).

[48] Fleming, G.G., Senzig, D.A., McCurdy, D.A., Roof, C.J., Rapoza, A.S., "Engine Installation Effects for Four Civil Transport Airplanes: Wallops Flight Facility Study," Cambridge, MA: Volpe Center, U.S. Department of Transportation, November 2003.

[49] Van Boven, M., "Lateral Attenuation of A300 Aircraft", Presentation at Meeting of SAE A-21 in Milan, Italy, June 2003.

[50] Storeheier, S.A., Randeberg, R.T., Granoien, I.L.N., Olsen, H., Ustad, A., "Aircraft Noise Measurements at Gardermoen Airport, 2001, Part 1: Summary of results," Trondheim, Norway: SINTEF Telecom and Informatics, June 2002.

[51] Smith, M.J.T., Ollerhead, J.B., Rhodes, D.P., White, S., Woodley, A.C., "Development of an Improved Lateral Attenuation Adjustment for the UK Aircraft Noise Contour Model, ANCON," London, England: Civil Aviation Authority, Draft, February 2002.

[52] Plotkin, K.J., Hobbs, C.M., Bradley, K.A., "Examination of the Lateral Attenuation of Aircraft Noise," Arlington, VA: Wyle Laboratories, July 1999.

[53] Embleton, Tony F.W., Piercy, Joe E., Daigle, Giles A., "Effective flow resistivity of ground surfaces determined by acoustical measurements," Journal of Acoustical Society of America, Vol. 74, No. 4, pp 1239-1243, 1983.

[54] Chessell, C.I., "Propagation of noise along a finite impedance boundary," Journal of Acoustical Society of America, Vol. 62, pp 825-834, 1977.

[55] Embleton, Tony F.W., Piercy, Joe E., Olson, N., "Outdoor Sound Propagation Over Ground of Finite Impedance," Journal of Acoustical Society of America, Vol. 59, No. 2, pp 267-277, February 1976.

[56] Chien, C.F., Soroka, W.W., "Sound Propagation Along an Impedance Plane," Journal of Acoustical Society of America, Vol. 43, No. 1, pp 9-22, 1975.

[57] Z. Maekawa, "Noise Reduction by Screens," Mem. Fac. Eng., Kobe University, Vol. 12, 1, 1966.

[58] Barry, T.M., Regan, J.A., "FHWA Highway Traffic Noise Prediction Model", Research Report FHWA-RD-77-108, Washington, D.C.: Federal Highway Administration, December 1978.

[59] "Statistical Analysis of FHWA Traffic Noise Data," Research Report FHWA-RD-77-19, Washington, D.C.: Federal Highway Administration, July 1978.

[60] http://wasmerconsulting.com/nmplot.htm, October 4, 2004.

[61] Technical Memo, <u>Addendum: Natural Ambient Sound Levels for use in Noise Modeling of Grand Canyon NP</u>, HMMH Job No. 295860.05, Harris Miller Miller & Hanson Inc., February 5, 1999.

[62] ISO 389-7:1998 Acoustics - Reference zero for the calibration of audiometric equipment - Part 7: Reference threshold of hearing under free-field and diffuse-field listening conditions, European Committee for Standardization.

[63] Green, David M. and Swets, John A., "Signal Detection Theory and Psychophysics." New York: John Wiley and Sons, Inc, 1966.

[64] Fidell, et. al., "Predicting Annoyance from Detectability of Low-Level Sounds," Journal of the Acoustical Society of America, 66(5), November 1979, p. 1427 – 1434.

[65] Sutherland, et. al., "Atmospheric Sound Propagation," Encyclopedia of Acoustics, Vol. 1, 1979, p. 351 – 353.

[66] "National Parks Air Tour Management Plan (ATMP) Program, Inputs to the FAA's Integrated Noise Model for Mount Rushmore National Memorial and Badlands National Park," U.S. DOT, Research and Special Programs Administration, John A. Volpe National Transportation Systems Center Acoustics Facility: Cambridge, MA, October 22, 2004.

Appendix A: FAA/NPS Letter to FICAN, Terms of Reference and Statement of Work

U.S. Department
of Transportation

**Federal Aviation
Administration**

U.S. Department
of Interior

SEP 2 2003

Mr. Alan Zusman
Department of the Navy
Naval Facilities Engineering Command
1322 Patterson Avenue SE Suite 1000
Washington Navy Yard, DC 20374-5065

Dear Mr. Zusman:

On behalf of the Federal Aviation Administration (FAA) and the Department of the Interior (DOI), we would appreciate the advice of the Federal Interagency Committee on Aviation Noise (FICAN) on some matters related to the measurement and assessment of the effects of aircraft noise due to overflights of units of the National Park System. We ask that you, as current FICAN chairman, convey our request to the Committee for their determination as to whether this would be an appropriate activity for FICAN.

The development of reasonable scientific methods that can be used for the assessment of the effects of aircraft noise in national parks characterized by low-level sound environments involves a number of technical issues related to the measurement, analysis, and characterization of sound. The FAA and the National Park Service (NPS) have made some progress in working cooperatively to reach consensus concerning how to address these issues under the National Environmental Policy Act of 1969 (NEPA) and the National Park Overflights Act of 1987 (NPOA), and the National Park Air Tour Management Act of 2000 (NPATMA). In light of recent rulings by the U.S. Court of Appeals for the D.C. Circuit (Grand Canyon Trust v. Federal Aviation Administration, 290 F.3d 339 (D.C. Cir. 2002 and United States Air Tour Association, et al., v. Federal Aviation Administration, 298 F.3d 997 (D.C. Cir. 2002) and requirements under the NPATMA, the FAA and the DOI have agreed that FICAN assistance on technical and scientific issues might be of mutual benefit. The DC Circuit rulings address requirements for assessing aircraft noise impacts under the NEPA and the NPOA. Section 808 of the NPATMA provides that "Any methodology adopted by a Federal agency to assess air tour noise in any unit of the national park system (including the Grand Canyon and Alaska) shall be based on reasonable scientific methodology."

Enclosed are the proposed terms of reference for the activity as discussed by our two agencies. Because of the urgency of the matter and its relation to a proposed alternative dispute resolution process currently underway, we are requesting a 2-year deadline on the tasks to be given to FICAN. In the interest of assisting FICAN and in recognition workload associated with the technical details of the proposed study, our agencies are ready to make available a team of contractors to support the Committee and to provide peer review. The agencies have agreed on a plan to work out the details to jointly fund the effort. Further, because of the dispute resolution process timetable we would also like FICAN to focus initially on the Grand Canyon issues described under item A.

The FAA and NPS representatives on FICAN, Tom Connor and Bill Schmidt, respectively, can provide further details and answer any questions that the Committee may have. Our representatives will also serve to report back on whether the FICAN accepts the assignment and terms of reference.

Enclosure

Sincerely,

Sharon L. Pinkerton
Assistant Administrator for Aviation Policy, Planning, and Environment

Paul Hoffman
Deputy Assistant Secretary for Fish and Wildlife and Parks

Terms of Reference

Aircraft Noise in National Parks

I. Scope

The Federal Interagency Committee on Aviation Noise (FICAN) shall assist the Federal Aviation Administration (FAA) and the National Park Service (NPS) with a review of the technical and scientific matters related to methodologies for assessing aircraft noise in units of the national park system. The review will be limited to existing information, with no conduct of new research.

II. Tasks

The joint FAA and NPS issues are conveyed as tasks to be undertaken as individual task orders, based on mutual needs and priorities of the two agencies. The tasks are to be assigned to one of 3 subject areas as follows:

A. <u>Review of Models for Assessing Noise in Grand Canyon National Park, as well as in Other National Parks.</u> Review the joint FAA-NPS Aircraft Noise Model Validation Study (See Aircraft Noise Model Validation Study, HMMH Report No. 295860.29, January 2003) and provide recommendations on the appropriate use and limitations of computer models and other tools for the calculation of aircraft noise in GCNP, and determine the extent to which this study may be helpful in other national parks (e.g., compression algorithm for overlapping of aircraft audibility when aircraft fly in close succession). The specific focus of this task will be on the most currently available version of applicable computer models, several of which have been updated since the January 2003 Model Validation Study.

B. <u>Review of Noise-Related Technical Issues for Grand Canyon National Park (GCNP), as well as for other National Parks.</u> FICAN shall provide technical advice and recommendations to the FAA, NPS, and other stakeholders with respect to noise in the Grand Canyon National Park, as well as in other parks. This task will be undertaken based on the mutual needs and priorities of the NPS and FAA. Specific work elements under this task might include the following:

1. Provide advice on a system for measuring substantial restoration of natural quiet at GCNP based upon the definition established by the NPS ("substantial restoration requires that 50 percent or more of the park achieve 'natural quiet' (i.e., no aircraft audible) for 75-100 percent of the day"). The advice should include but is not limited to the choice of "day" as used in the definition of substantial restoration (e.g., average day vs. peak day vs. peak season average day) in consideration of the decision of the U.S. Court of Appeals for the District of Columbia, U.S. Air Tour Ass'n. v. FAA 298 F.3d 997 (D.C. Cir. 2002), and the determination of the ambient sound environment, (e.g., the process used to gather data to represent ambient).

2. Provide recommendations concerning use of aircraft audibility to quantify natural quiet and examine the use of different thresholds to quantify audibility of aircraft noise as part of two-zone system for evaluating achievement of natural quiet (see *Change in*

Noise Evaluation Methodology for Air Tour Operations Over Grand Canyon National Park, 64 FR 3969). The examination should include but is not limited to quantifying audibility (e.g., detectibility vs. noticeability) and the scientific basis of a zoning system (e.g., the relationships among audibility thresholds and the differing park uses, differing levels of park resource protection, and differing levels of development to serve park visitors).

3. Provide a review and recommendations regarding the use of aircraft noise certification data for assessment of audibility. Compare noise certification data with field-measured data, assess the correlation between the two, and recommend appropriate uses or limitations of both the certification data and the field measured data.

4. Provide a review and assessment of methods used in the U.S. and elsewhere to assess the impacts of noise, particularly aircraft noise, on resources of and visitors to protected sites such as national parks. Included in this review shall be the guidance documents used or proposed for use by authorities controlling airspace use and by those responsible for protecting resources and the quality of visitor experiences relative to noise intrusions.

5. Provide a review and detailed assessment of the methodologies used by agencies responsible for controlling the generation of aircraft noise and those responsible for protecting resources to comply with the cumulative impact provisions of NEPA. Included in the review shall be an assessment of the successful defense of those methodologies.

6. Provide a review and an assessment of the relative effectiveness of alternative measures for mitigating noise in units of the National Park System. The review shall include, to the extent feasible, an assessment of new technologies proposed for noise mitigation.

C. Soundscape within the units of the National Park System (See NPS Director's Order #47: Soundscape Preservation and Noise Management, 12/1/2000). This task will be undertaken based on the mutual needs and priorities of the NPS and FAA. Specific work elements under this task might include the following:

1. Provide a review, assessment, and recommendations on methods used to measure ambient sound conditions in low-sound level settings such as are found in the National Parks. This includes, but is not limited to, the equipment, the statistical treatments of the data, guidance documents used or proposed by both agencies, and the various metrics used to describe the ambient in these settings (e.g., L_{90} vs. L_{50}).

2. Examine representative data on the variety of civil, commercial, public, and military aircraft operations that may fly over or near units of the national park system and provide recommendations on methodologies for determining the scope and extent of aircraft noise intrusions. The examination should include but is not limited to the adequacy of the scientific basis of the data and methodologies for use in aircraft noise analysis and the assessment of the extent that intrusions become impacts.

3. Provide advice on the use of the best science available to determine the impact of existing or proposed aircraft noise sources on the soundscape, wildlife, aquatic and marine life, cultural resources, other resources and values, and the visitor experience, as appropriate. The advice should address but is not limited to the determination of the

affected environment and the type, magnitude, duration, and frequency of occurrence of aircraft noise that may be compatible or incompatible with national park environments and purposes (Reference relevant NPS laws, regulations, and policies).

III.   Schedule

The FICAN shall produce a draft report for Task A no later than 3 months after task initiation. The contractor shall provide a final report on Task A to FICAN no later than 30 days after receipt of comments on the draft final report. For Tasks B and C, a schedule will be mutually agreed to by FICAN and the two agencies, based on specific work items to be undertaken.

STATEMENT OF WORK

Background:

The Federal Interagency Committee on Aviation Noise (FICAN) began in 1993 as a technical liaison among agencies to develop recommendations and priorities on needed research and noise assessment issues. FICAN is assisting the Federal Aviation Administration (FAA) and the National Park Service (NPS), two of its member agencies, in addressing technical issues related to the implementation of the National Parks Overflights Act of 1987 for the Grand Canyon National Park and the Air Tour Management Act of 2000.

The Congressional mandates relevant to this activity are as follows:

- The National Parks Overflights Act of 1987, which required the Department of the Interior (DOI) to submit to the Congress recommendations to protect the natural resources of Grand Canyon National Park (GCNP) from the adverse impacts of aircraft overflights. Specifically, the NPS was required to provide for "the restoration of natural quiet and experience" in the GCNP, prohibit most flights below the rim of the Canyon, and designate flight-free zones. The FAA has the obligation under that same Act to assess aviation safety relative to the Interior Secretary's recommendations regarding proper minimum altitude of aircraft flying over units of the National Park System, and to notify the Secretary of any adverse effects, which the implementation of such recommendations would have on safety of aircraft operations.
- The National Parks Air Tour Management Act of 2000, which directed the FAA, with the cooperation of the NPS, to develop Air Tour Management Plans (ATMPs) for all National Parks with commercial air tours except for Grand Canyon NP, Tribal lands within or abutting Grand Canyon NP, flights transiting Lake Mead NRA, and the parks in Alaska. The FAA has the authority to preserve, protect, and enhance the environment by minimizing, mitigating, or preventing the adverse effects of aircraft overflights on public and tribal lands. The objective of the ATMPs is to develop acceptable and effective measures to mitigate or prevent significant adverse impacts from the air tours on the natural and cultural resources, visitor experiences, and tribal lands. The FAA is also obligated to designate reasonably achievable requirements for fixed wing and helicopter aircraft necessary for such aircraft to be considered as employing quiet aircraft technology.
- The National Parks Service Organic Act, which called for the promotion and regulation of the use of the Federal areas known as national parks, monuments, and reservations in order to conserve the scenery and the natural and historic objects and the wildlife therein and to provide for the enjoyment of the same in such manner and by such means as will leave them unimpaired for the enjoyment of future generations.

Task:

The scope of work, as defined in Task A of the Terms of Reference Document (See attached *Terms of Reference - Aircraft Noise in National Parks*) and incorporated wholly in this Statement of Work, shall be limited to the review of existing data and other information for the tasks described below as directed by FICAN (if additional research is recommended by FICAN, the Contractor(s) will provide additional technical support as directed by the FAA). Per FICAN approval, subject area experts will be contracted by the Contractor(s), as needed. The Contractor(s) will support FICAN in the following task:

Task A: Conduct a comprehensive review and assessment of available computer models to be used for assessing noise in Grand Canyon National Park, as well as in other National Parks

1. Review the joint FAA-NPS Aircraft Noise Model Validation Study (See *Aircraft Noise Model Validation Study, HMMH Report No. 295860.29, January 2003*) and provide recommendations on the appropriate use and limitations of computer models and other tools for the calculation of aircraft noise in GCNP, as well as in other National Parks.

   a) Identify the acoustical limitations of available computer models in predicting noise in GCNP, as well as in other National Parks. For example:

      i) The inability of some models to account for terrain shielding, dense foliage effects, and/or propagation over mixed terrain types, e.g., hard/soft ground.

      ii) The inability of some models to account for overflights, which occur at the same time or overlap slightly.

      iii) The use of contouring algorithms for noise prediction at specific locations.

      iv) The effect of source directivity within each model (e.g., some models incorporate a 3-dimensional source directivity for a limited number of aircraft into calculations. The limitations associated with including aircraft for which only 2-dimensional source directivity data are available should be studied).

   b) Identify other practical limitations of available computer models in predicting noise in GCNP, as well as in other National Parks. For example:

      i) A reasonableness comparison of model run-times.

      ii) The input data requirements for each model.

      iii) The availability and coverage of aircraft source noise data (and estimated resources to correct shortcomings in each model).

      iv) A comparison of model output, including available noise metrics and analysis capabilities.

      v) A comparison of model availability, usability and documentation.

   c) Provide recommendations on the appropriateness of computer models and other tools for the calculation of aircraft noise in GCNP, as well as in other National Parks.

   d) Assess the current usability of available computer models and other tools for the calculation of aircraft noise in GCNP, as well as in other National Parks.

Assess the extent of peer review of the methodologies employed in computer models used to predict noise in GCNP, as well as in other National Parks.

Schedule and Deliverables:

The contractor shall produce a draft final report for Task A no later than 3 months after task initiation.

The draft final report of the findings and recommendations shall include: (1) technical issues that are insufficiently addressed or not addressed by the findings/ recommendations; (2) advice for resolving unaddressed technical issues; and (3) recommendations for technically defensible alternative approaches. If additional research is recommended by FICAN, the Contractor will provide additional technical support as mutually agreed to by the FAA and NPS. The contractor shall provide a final report on Task A to FICAN no later than 30 days after receipt of comments on the draft final report.

Appendix B: Summary of Enhancements Included in INM Version 6.2 for use in the National Parks

B.1 Summary of INM 6.2 Updates

The Federal Aviation Administration, in cooperation with other agencies, has been engaged in research activities designed to improve noise modeling for aviation projects that require environmental noise analysis and disclosure. The majority of this research is performed under the Society of Automotive Engineers Aircraft Noise Committee (SAE A-21). These activities are closely coordinated with similar groups within the European Civil Aviation Conference (ECAC) and the International Civil Aviation Organization (ICAO). INM 6.2 includes several changes related to aircraft noise/performance for commercial aircraft and the modeling of aviation noise over national parks. The new acoustic modeling procedures related to audibility are outside the applicability of the core SAE, ECAC and ICAO modeling documents. They are developed to address conditions in national parks, which may contain very low-level ambient sound level conditions.

B.2 Commercial Aircraft Noise/Performance Database

Review of the core INM noise and performance database has shown that certain aircraft have grown in maximum allowable takeoff weight, operating range and thrust setting since the database was developed in the late 1980's. This release of INM updates five aircraft types to better reflect the current "in-service" fleet. The INM 757PW, 757RR and, 777200 have been updated to reflect growth in maximum allowable takeoff weight and engine thrust since data for these aircraft were produced for previous versions of INM. The 747400 and the 737700 have been re-derived to follow the same rules as used for current aircraft (See Attachment 1). While individual stage weights have been reduced for the 747400, the overall range of operating weights remains the same and users still retain the capability to model the 747400 over its entire operating range, including the maximum allowable takeoff weight. Attachment 1 provides general overview of the guidelines FAA and Eurocontrol have been developing to harmonize model data across manufacturers.

B.3 Noise Modeling for National Parks

In 1996, the Federal Aviation Administration began conducting studies to assess Special Flight Rules in the vicinity of Grand Canyon National Park. These studies provided noise disclosure assessments under NEPA that required noise models capable of evaluating a broad area for both fixed-wing and helicopter operations, using metrics not contained in the standard release of INM. Since 1996, studies requiring these capabilities have continued with FAA analysis performed using special research or application specific versions of the INM. The FAA has updated its current release of INM to make these modeling capabilities publicly available. Primary updates to INM to support national parks modeling include an expansion of the noise/performance database to include more general aviation aircraft and the inclusion of supplemental metrics that have been used in analysis of Grand Canyon National Park. Validation studies have highlighted the need to further enhance INM to include more detailed terrain data and the ability to model line-of-sight blockage. A more detailed description of these enhancements is included in the release notes below.

The FAA's immediate use for INM 6.2 is to complete the aircraft overflight noise analysis of Grand Canyon National Park to assess the substantial restoration of natural quiet, as required by US law. The National Park Service (NPS) is particularly interested in the ability to calculate Time Audible for national parks. The FAA has not established a preferred supplemental metric or metrics for national park noise analysis, and these enhanced INM capabilities should not be presumed to constitute an endorsement of these particular metrics over others. Neither should aircraft audibility, when calculated, be presumed to be a measure of an adverse or significant impact. The FAA, in consultation with the NPS, will advise on the use of metrics for national park noise analysis on a case-by-case basis until standardization is achieved based on further technical and scientific review.

B.3.1 New Noise Metrics

Two new noise metrics have been added to this public release of INM: Time Audible (TAUD), which is the amount of time that aircraft are audible, and Change in Exposure (delta dose or DDOSE), the change in noise exposure associated with aircraft operations (i.e., the arithmetic difference between aircraft noise exposure and ambient sound level). The user also has the ability to calculate the Percentage Time Audible (%TAUD) for a specific time period, such as 24 hours, or a shorter period representing a time in which an area may be subjected to aircraft overflights. These metrics have been used in US studies related to National Parks and are made publicly available with the INM 6.2 release. As mentioned above, their inclusion should not be presumed to constitute an endorsement of these particular metrics.

B.3.2 Using TAUD and DDOSE in INM

TAUD and DDOSE have been used in research versions and are now merged into the INM 6.2 release series in way that does not require modification to the underlying database that contains the inputs to INM studies. No special conversion software is necessary. The TAUD and DDOSE metrics are selected through the Noise Metric drop-down list in the Run-Options window for each INM Case. They are not available for selection under the Grid section of the Run Options window and therefore gird point values can only be calculated for one of the two new metrics during each CASE run. In other words, if a user wanted to run TAUD and DDOSE for a CASE called BASE_2005, they would need to run the case twice selecting the appropriate metrics or create two cases called BASE_2005_TAUD and BASE_2005_DDOSE.

Upon selecting TAUD or DDOSE from the Noise Metric drop down list box, the user may run INM to produce results using these metrics. TAUD and DDOSE values can be produced for Standard and/or Detailed grids. The values will appear in the Metric column of the Standard Grid report. The two metrics will be treated as any other metric in the detailed grid report. In summary, TAUD and DDOSE behave just like ordinary <u>user-defined metrics</u> that have been created to be part of the INM 6.1 release series.

Audibility compares aircraft noise against background noise to determine if noise may be detected. The process is based on detectability theory along with research that has assessed human detectability under different environments. *Attachment 3 – Calculating Audibility* provides additional specifics on the theory and background. Audibility requires highly detailed inputs and results may be very sensitive to the quality of input data. Guidance on developing these inputs (i.e. an ambient map file) is still in progress and subject to further scientific review. Accounting for background noise requires additional input into INM and the INM 6.0 User's Guide address the mechanics of importing an ambient map file. The specifics on using TAUD and DDOSE are given below.

Relative Threshold Audibility

Selecting "Relative Threshold" within Grid Setup allows the user to calculate audibility based on spectral ambient data. An ambient file was first introduced in INM 6.0 to support a metric for time above an ambient level (TALA) and it is described in Section 10.1 of the INM 6.0 User's Guide. This file may still be used to support TALA calculations. For TAUD, the user will need a modified version of this file accompanied by a second file, which contains 1/3-octave spectral information mapped to a cumulative A-weighted sound level. Percentage of park area is determined by a geographic boundary file, which gives the official demarcation of park boundaries. A description of this modified *ambient.txt* file, and the spectral file called *ambi_map.txt* is given in ***Attachment 2 – Ambient Data Input Files***. The ambient grid file is a text grid file, which assigns a number, representing the A-weighted ambient sound level, to study area grid points. The ambient spectral map file correlates unique, A-weighted spectra to the ambient sound levels specified in the ambient grid file. The geographic boundary file *boundary.txt* uses the same format as the Polyline TXT file described in Section 3.5.2 of the INM 6.0 User's Guide. The geographic boundary can be easily imported into INM for viewing in the Input Graphics or Output Graphics windows using the Import Polyline TXT file function, also described in Section 3.5.2 of the User's Guide.

For any case in which a user-defined one-third octave band spectral level is below the Equivalent Auditory System Noise (EASN) floor, the INM will replace (mask) that level using the associated EASN level so as to not predict an unreasonably high audibility level. If any error or warning messages are produced by the INM associated with the ambient sound level data, the INM will alert the user and detailed information may be found in the *ambient_error.txt* and/or *ambient_warning.txt* files, respectively, written to the case directory. Typical errors are usually data anomalies which would likely produce erroneous results; INM therefore aborts processing. A typical error might include an A-weighted sound level in the ambient grid file, which does not have an associated spectrum in the spectral map file. Warnings, on the other hand, are issues which INM identifies as potentially but not necessarily erroneous. For example, a warning is produced if the A-weighted value specified for a given spectrum in the ambient spectral map file does not match the A-weighted sum of the spectral data. In this case, the INM does continue to perform calculations and a warning file is produced.

Fixed Threshold Audibility (Screening Analysis)

Determining audibility <u>requires</u> the use of input data containing information on ambient sound levels and is described in the <u>relative threshold audibility</u> section above. A user may perform a screening analysis, which may aid in prioritizing the collection of ambient data, which is necessary for the full analysis. This screening process models audibility conservatively and assumes no ambient noise or ambient levels higher than the EASN threshold. In lieu of actual spectral ambient data, use of the EASN threshold is considered to be the most conservative assumption. In practice, the actual amount of time aircraft are audible is likely to be less than predicted using this conservative screening assumption. The EASN threshold is presented in both a figure and in tabular format at the end of ***Attachment 3 – Calculating Audibility***.

Percent Time Audibility

Using either the Fixed Threshold or Relative Threshold options outlined above, the user may calculate the percentage of time that aircraft are audible; this function is enabled by selecting the "Do Percent of Time" check box in the Grid Setup window and entering the duration of time over which to calculate the percentage.

INM audibility calculations do not directly account for overlapping aircraft operations. If all or a portion of the audibility of two unique aircraft overlap in time the model may tend to overpredict audibility. For this reason percent time audibility is capped at 100 percent.

Change in Exposure

Change in Exposure may be modeled in INM by selecting the DDOSE noise metric. DDOSE is defined as the arithmetic difference between aircraft noise exposure and ambient sound level. Similar to TAUD, the user has multiple means for calculating DDOSE. Specifically, there are several options for selecting the ambient sound level, which is utilized in the calculation of Change in Exposure.

Fixed Threshold Change in Exposure

Selecting "Fixed Threshold" within Grid Setup allows the user to calculate the Change in Exposure relative to a fixed, user-defined value. Unless normalized to another time period (see below), the metric utilizes a 12-hour equivalent sound level. If the Fixed Threshold box is selected, a box appears next to the option labeled "Fixed threshold (dB)" in which the user may enter the actual threshold value.

Relative Threshold Change in Exposure

Selecting "Relative Threshold" within Grid Setup allows the user to calculate Change in Exposure based on A-weighted ambient data. The calculation requires the ambient grid file highlighted above. Change in Exposure may be calculated in reference to the values in the ambient grid file as is, or with an absolute delta applied to the ambient data. This delta value may be entered into the box labeled "Ambient + Delta (dB)".

Change in Exposure Normalized to other Time Periods

Using either the Fixed Threshold or Relative Threshold options outlined above, the user may calculate the Change in Exposure normalized to a time other than 12 hours; this function is enabled by selecting the "Do Percent of Time" check box in the Grid Setup window and entering the duration of time over which to calculate the percentage.

B.3.3. National Parks Noise Database Enhancements

Four general aviation aircraft have been added to the INM database: Piper PA28-161 Warrior, Piper PA30 Twin Comanche, Piper PA31-350 Navajo Chieftain, and Maule M-7-235. Data for these aircraft have been developed according to procedures outlined in SAE-AIR-1845 that define the aircraft noise source and relate this source to the changing power state of the aircraft.

Data for two helicopters have been added to the INM supplemental database for helicopters. Instructions for using this data in INM are provided in the INM 6.0c release notes. New helicopter data include the Eurocopter EC-130 and Robinson R-22. Data for the four GA aircraft and the two helicopters are derived from a flight test conducted during 2002 at Fitchburg Municipal Airport in Fitchburg, MA.[8]

The general aviation aircraft and helicopters added to the INM database were included to aide in the modeling of aircraft noise for the joint FAA-NPS Air Tour Management Plan (ATMP) program. In support of the ATMP program, a database containing Interim Operating Authority (IOA) applications is maintained by FAA AWP-4. Version 3A of this database, dated June 3, 2003, was used to compile a list of anticipated ATMP fleet operations coverage for the INM. The additional aircraft in the INM database have been added given the current list of National Parks scheduled for ATMPs and the associated air tour operations at those parks.

B.3.4. Terrain Modeling - Line-of-Sight Blockage

The capability to account for line-of-sight (LOS) blockage has been added to INM Version 6.2. This feature accounts for the added attenuation due to LOS blockage from terrain features. LOS blockage may be implemented utilizing the same 3CD terrain data already in use by the model to correct source-to-receiver distance for terrain elevation; it may also be implemented using the additional terrain data types outlined below. The LOS blockage calculation is based on the difference in propagation path length between direct LOS propagation and propagation over the top of terrain features. The path length difference is used to compute the Fresnel Number (N_0), which is a dimensionless value used in predicting the attenuation provided by a noise barrier positioned between a source and a receiver. Figure B-1 illustrates LOS blockage from a terrain feature. The formula used by INM to calculate Fresnel Number is given in References 57 and 58.

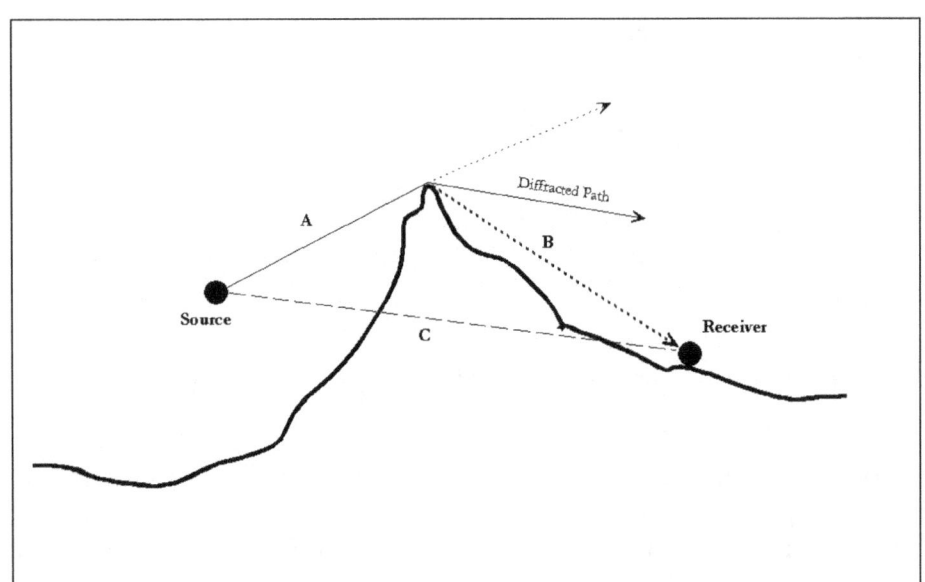

Figure B-1. Line-of-Sight (LOS) Blockage Concept

LOS blockage may be invoked by checking the "Do Line-of-Sight Blockage" check box in the *Run Options* window, after having selected the *Do Terrain* option.

In order to calculate LOS blockage, the INM requires terrain data for an area that covers the extent of all of the desired output grid points, including the calculated flight paths. Standardized terrain data sets often do not contain data for areas extending far out into large bodies of water. Therefore, if any of the calculated flight paths for an INM study extend far out over a large body of water, there may not be enough terrain data available to meet the INM's terrain data coverage requirements for LOS blockage. A process for automatically filling in terrain data in these situations is currently being developed and will be incorporated into INM 6.2 prior to the full public release. For the current Beta release of INM 6.2, however, the user may use the LOS blockage capability on INM studies covering airports near very large bodies of water, such as the ocean, by creating terrain data files spanning the water areas, which contain appropriate altitude data.

B.3.5. Terrain Modeling – Additional Terrain Data Capability

INM 6.2 has been expanded to include the use of higher resolution terrain data than previous versions of INM. Specifically, INM will now utilize GridFloat and Digital Elevation Model (DEM) data. Both types of data are maintained as a part of the National Elevation Dataset (NED) by the U.S. Geological Survey (USGS). This data may be downloaded free of charge from the USGS National Map Seamless Data Distribution System (http://seamless.usgs.gov). The data may also be purchased on CDROM media for a nominal fee. When downloading or ordering GridFloat data from this site, a user must save to the NED GridFloat, text format.

GridFloat and DEM data may be imported for viewing in the Output Graphics window using the Terrain Processor found in File // Import Data into Study // Terrain Files. The process for importing the two new terrain formats is the same as the process for importing 3CD/3TX terrain data and is explained in the INM 6.0 Users' Guide. The noise calculation program uses the terrain data located in the Terrain Files Directory directly whereas the above import utility creates a special file for viewing Output graphics. As these are different files, it is up to the user to update the terrain data displayed in the Output Graphics to match the terrain data used to calculate the noise levels. INM will not do this automatically.

Imported terrain contours for all three terrain data types are stored under a new file name with a new file format in INM 6.2. The new file name is _terrain62.bin, and the new file format allows INM 6.2 to display larger amounts of contour data than was possible with previous versions of the INM. If an older INM study containing terrain contours in the old format is opened with INM 6.2, the data in the old _terrain.bin file will automatically be moved to the new _terrain62.bin file and the old _terrain.bin file will be deleted. Similar to use of 3CD/3TX data, all required files must be placed in the directory specified in the *Terrain Files* option of the *File Locations* dialog box. For GridFloat data, this includes files with the *.flt* (terrain elevation data), *.hdr* (metadata including boundaries) and, *.prj* (data projection information including datum) file extensions. DEM data must include files with the *.dem* file extensions.

Unlike 3CD/3TX, the GridFloat data format is non-proprietary and stores its data in a latitude-longitude coordinate system that can be used worldwide. GIS systems such as ArcInfo have the option of saving data to this format and there are sources for GridFloat data for areas outside the United States. Note that ArcInfo will not add the *.flt* file extension to the GridFloat file containing the terrain data. If using ArcInfo, this filename will need to be modified by the user to have the *.flt* extension.

B.3.6. Disabling Lateral Attenuation for Propeller Aircraft

Lateral attenuation in INM is based on the draft update to SAE-AIR-1751, Prediction Method for Lateral Attenuation of Airplane Noise During Takeoff and Landing. This document was developed principally for commercial jet aircraft. Military aircraft and helicopters are not addressed by this document. Consequently, INM employs different lateral attenuation equations depending on the class of aircraft. An aircraft's class is determined by its spectral class assignment. A complete description of these assignments is given in Attachment 5.

SAE A-21 is currently undertaking an update to this AIR. It is anticipated that a future update to the document will incorporate the capability to model propagation over acoustically hard surfaces such as water or rocks. The capability to turn off lateral attenuation for helicopter and *propeller aircraft* has been added to INM Version 6.2. This feature simulates propagation over acoustically hard ground. It may be useful for national parks with a significant amount of hard, rock face surfaces.

B.3.7. Level Flyover NPD Curves

Level flyover NPD data have been added for several propeller-driven aircraft in INM 6.2 because level flyover operations constitute a significant amount of the overall operations over national parks. These data can only be accessed by creating user-defined fixed-point profiles for these aircraft. In a fixed-point profile the data are accessed by specifying the Flyover/Afterburner Operational Mode for each applicable profile step. Doing so will limit the thrust settings available for use with those profile steps to those thrust settings identified in the flyover NPD data sets.

When using the Flyover/Afterburner Operational Mode to access the flyover NPD data, it is important to account for the way the INM handles thrust changes in the different operational modes. For example, when the thrust setting changes between two profile steps using the Approach or Depart Operational Modes, the INM transitions the thrust between the two values over the entire profile segment length. When the thrust setting changes between two profile steps, and one or both of those steps use the Flyover/Afterburner operational mode, the INM handles the thrust transitions differently. The INM adds a new 100 ft segment to the profile and transitions the thrust between the two values over this short segment. For this reason it is recommended that the flyover NPD data only be used in conjunction with overflight profiles and that those overflight profiles utilize the flyover data throughout the entire profile.

B.4. New MapInfo Interchange File Export Function

The INM 6.2 software release contains a new export function that writes MapInfo Data Interchange Format files containing INM graphics output layers. These files can be read by MapInfo Professional and other GIS programs that support the MapInfo Data Interchange format. The "File // Export As MIF/MID" function is available when the Output Graphics window is active. Operation of the function is similar to the "File // Export As Shapefile" function.

Output Graphic layers that are enabled (visible) are exported. Data associated with items visible in Output Graphics, such as population numbers at population points or noise levels at standard grid points, are also exported. Two pre-named files for each active graphics layer (*.mif and *.mid) are written to an existing directory that is selected by the user. For example, noise contours are exported to *Noise-Contours.mif* and *Noise-Contours.mid* files. A prefix can be added to all file names to help differentiate between different sets of MapInfo files. Coordinates can only be exported as latitude/longitude decimal degrees. The table below lists the 11 MapInfo Data Interchange files that are available.

MapInfo Files (.mif, .mid)
Airport-Drawings.*
Airport-Runways.*
Flight-Tracks.*
Grid-Points.*
Locations-Points.*
Noise-Contours.*
Overlay-Contours.*
Population-Points.*
Radar-Tracks.*
Terrain-Contours.*
Tiger-Lines.*

B.5. Database Modifications

1. Data for the Boeing 757-200 with PW2037 engines has been updated for INM 6.2. The existing INM identifier is 757PW and the noise identifier is PW2037. This aircraft reflects a growth in maximum allowable takeoff weight from 240,000 to 255,000 and new weight-to-stage length rules that are increased from assumptions made in the late 1980's. There are three sets of procedural departure profiles: ICAO_A, ICAO_B, and STANDARD all of which have stage lengths 1 through 7 with weight 7 being the maximum takeoff weight. Noise-Power-Distance data has been updated. Though similar, there are now more curves for approach conditions, and the aircraft now has LAMAX and PNLTMAX curves developed uniquely for this airframe/engine variant. New high temperature jet thrust coefficients have been added for modeling aircraft performance above engine break point temperatures.

2. Data for the Boeing 757-200 with RB211-535E4 engines has been updated for INM 6.2. The existing INM identifier is 757RR and the noise identifier is RR535E. This aircraft reflects a growth in maximum allowable takeoff weight from 220,000 to 255,000 and new weight-to-stage length rules that are increased from assumptions made in the late 1980's. There are three sets of procedural departure profiles: ICAO_A, ICAO_B, and STANDARD all of which have stage lengths 1 through 7 with weight 7 being the maximum takeoff weight. Noise-Power-Distance data has been updated. The NPD for this aircraft was updated for INM 6.1. This release normalized this NPD to an ICAO atmosphere. New high temperature jet thrust coefficients have been added for modeling aircraft performance above engine break point temperatures.

3. Data for the Boeing 737-700 with CFM56-7B engines has been updated for INM 6.2. The existing INM identifier is 737700 and the noise identifier is CF567B. The engine type has been updated to CFM56-7B24. This aircraft reflects a growth in maximum allowable landing weight from 138,000 to 129,200 and new weight-to-stage length guidelines that are increased from previous assumptions. There are three sets of procedural departure profiles: ICAO_A, ICAO_B and STANDARD all of which have stage lengths 1 through 6 with 6 being the maximum takeoff weight.

4. Data for the Boeing 777-200 with GE90-90B engines has been updated for INM 6.2. The existing INM identifier is 777200 and the noise identifier has been updated to the GE90. This aircraft reflects a growth in maximum takeoff weight from 535,000 to 656,000 pounds and available engine thrust from 77,000 to 90,000 pounds. The approach profile has been modified from a 1500 level flight segment to a 3000-foot level flight segment to make it consistent with other INM submissions and reflects a growth in maximum landing weight from 445,000 to 470,000 pounds. There are three sets of procedural departure profiles: ICAO_A, ICAO_B and STANDARD all of which have stage lengths 1 through 9 with 9 being the maximum takeoff weight. New high temperature jet thrust coefficients have been added for modeling aircraft performance above engine break point temperatures.

5. Data for the Boeing 747-400 with PW4056 engines has been updated for INM 6.2. There are three sets of procedural departure profiles: ICAO_A, ICAO_B and STANDARD all of which have stage lengths 1 through 9 with 9 being the maximum takeoff weight. Noise-Power-Distance data has been updated to include maximum level metrics. New high temperature jet thrust coefficients have been added for modeling aircraft performance above engine break point temperatures.

6. Data for the Piper PA28-161 Warrior were added to the INM database. The aircraft identifier is PA28 and the noise identifier is O320D3. Noise identifier O320D3 includes NPD data for three different RPM power settings over three different aircraft states (approach, departure and level-flyover). The STANDARD approach and departure profiles are both fixed-point profiles. A fixed-point overflight profile identified as LEVEL that uses the Flyover/Afterburner operational mode is also included in the database for this aircraft. Fixed-point profiles are used because research demonstrated that engine RPM provided the best correlation between aircraft state and aircraft noise source.

7. The PA28WA substitution is still in the *acft_sub.dbf*, and it is now equated to the new PA28. The new PA28 aircraft should be used in all INM studies, and users should take steps to change references to the PA28WA substitution to the new PA28 standard identifier. <u>A future version of INM will remove the PA28WA record from the *acft_sub.dbf* file, and a study using the PA28WA will have to be manually converted by the user.</u>

8. Data for the Piper PA30 Twin Comanche were added to the INM database. The aircraft identifier is PA30 and the noise identifier is IO320B. Noise identifier IO320B includes level flyover NPD data for three different thrust settings. The STANDARD approach and departure profiles are both procedural profiles. A fixed-point overflight profile identified as LEVEL that uses the Flyover/Afterburner operational mode is also included in the database for this aircraft. Because the SAE-AIR-1845 propeller performance equations do not account for the performance decrease with altitude in normally-aspirated piston engines, the INM's two thrust levels (MaxTakeoff and MaxClimb) were used to simulate the performance decrease: MaxTakeoff modeled full available power climb from Sea Level to 3000 feet and MaxClimb modeled full available power climb from 3000 feet to 10,000 feet.

9. The PA30 substitution was removed from the *acft_sub.dbf* file. INM 6.2 automatically converts the PA30 substitution, if used in a study, into the new PA30 aircraft.

10. Data for the Piper PA31 Navajo were added to the INM database. The aircraft identifier is PA31 and the noise identifier is TIO542. Noise identifier TIO542 includes level flyover NPD data for three different thrust settings. The STANDARD approach and departure profiles are both fixed-point profiles. A fixed-point overflight profile identified as LEVEL that uses the Flyover/Afterburner operational mode is also included in the database for this aircraft. The power parameter used in the profile points and the NPD curves is engine RPM.

11. The PA31 substitution was removed from the *acft_sub.dbf* file. INM 6.2 automatically converts the PA31 substitution, if used in a study, into the new PA31 aircraft.

12. Data for the Raytheon Beech 1900D were added to the INM database. The aircraft identifier is 1900D and the noise identifier is PT6A67. The STANDARD approach and departure profiles are both procedural profiles. This aircraft has two departure stage lengths. Measurements undertaken to derive the data showed that due to high frequency noise components in cruise condition, this aircraft has higher sound levels in cruise than at takeoff. Two departure NPD curves with different power settings, yet identical sound levels, have been added to eliminate possible problems when extrapolating outside the measured NPD range.

13. The BEC190 substitution is still in the *acft_sub.dbf*, and it is now equated to the new 1900D. The new 1900D aircraft should be used in all INM studies, and users should take steps to change references to the BEC190 substitution to the new 1900D standard identifier. <u>A future version of INM will remove the BEC190 record from the *acft_sub.dbf* file, and a study using the BEC190 will have to be manually converted by the user.</u>

14. Data for the Maule M-7-235 were added to the INM database. The aircraft identifier is M7-235 and the noise identifier is IO540W. No performance information or standard Approach and

Departure profiles for the Maule M-7-235 are included in this release. There is however a fixed-point overflight profile identified as LEVEL that uses the Flyover/Afterburner operational mode included in the database for this aircraft. The Maule NPD data are only intended to model level flyovers, similar to air tours over parklands. As with the Warrior and the Navajo Chieftain, the NPD power parameter is engine RPM. The higher engine RPM NPD represents the Maule's normal airspeed cruise flight. The lower engine RPM NPD represents the Maule's low speed level flight capability, typical of what might be flown over scenic areas within a park. When modeling this aircraft, users should only model overflight operations, as no departure or arrival information is provided.

15. The thrust setting types for all military aircraft have been changed to "other" ("X" in the nois_grp.dbf file) in the INM database. Previously some of the thrust setting types were incorrectly set to "percent".

16. Data for the Eurocopter EC-130 are now available for use in the INM. These data have been added to the npd_curv.dbf file located in the Helo\HeloExample INM subdirectory. Instructions for the use of the EC-130 data may be found in the Helicopter.doc file, originally disseminated with INM 6.0c and located in the Helo subdirectory of all subsequent versions of the INM.

17. Data for the Robinson R-22 are now available for use in the INM. These data have been added to the npd_curv.dbf file located in the Helo\HeloExample INM subdirectory. Instructions for the use of the R-22 data may be found in the Helicopter.doc file, originally disseminated with INM 6.0c and located in the Helo subdirectory of all subsequent versions of the INM.

18. The profile weights of the approach and departure NoiseMap profiles for the F16A aircraft have been changed from 90,000 lbs and 85,000 lbs, respectively, to a more realistic value of 33,000 lbs. This change has no effect on the actual profiles or the output of INM because the F16A uses only fixed-point profiles that are calculated independently of aircraft weight.

19. The initial speed for all standard fixed-point departure profiles has been changed from 35 knots to 0 knots in the INM database. This change has no effect on the flight paths or noise levels calculated for these profiles.

20. Duplicate thrust coefficient ID's and corresponding data for the GIIB and GIV aircraft have been removed from the INM database.

21. Two new noise metric identifiers, TAUD (Time Audible) and DDOSE (Delta Dose), have been added to the INM database.

22. Weight category assignments have been corrected for four aircraft in the INM database. The weight category assignments for the CNA55B and DOMIN aircraft have been changed from 'Small' to 'Large'. The weight category assignment for the 767400 aircraft has been changed from 'Large' to 'Heavy', and the assignment for the VULCAN aircraft has been changed from 'Heavy' to 'Large'. These changes have no effect on INM studies using these aircraft.

23. The engine type listed for the F18EF aircraft has been corrected from the F404-GE-400 engine to the F414-GE-400 engine. This change has no effect on INM studies using this aircraft, the correction only fixes a typo in the database rather than changing the actual engine and thereby changing the aircraft's noise data.

24. The standard fixed-point approach profiles for the 737800 and 757300 have been modified. The length of one of the thrust transition segments has been changed from 0 ft to 100 ft for each profile. These changes should have no significant impact on noise levels calculated for aircraft operations using these profiles.

B.6. Program Modifications

1. The ability to calculate the time audible (TAud) noise metric has been added to the program. TAud is defined as the time that aircraft are audible to an attentive listener. Calculation of TAud requires source and ambient sound level spectra for a given analysis location. TAud has been added to the Single-Metric type of noise metrics only; it is not available for modeling Multi-Metrics.

2. The ability to calculate the change in exposure (delta dose or DDOSE) noise metric has been added to the program. DDOSE has been added to the Single-Metric type of noise metrics only; it is not available for modeling Multi-Metrics.

3. The ability to disable the use of ground-to-ground lateral attenuation when calculating noise generated by helicopter and propeller-driven aircraft has been added to the program. Lateral attenuation can be turned off for these aircraft by selecting "No-Prop-Attenuation" in the new Lateral Attenuation drop-down list located in the Run // Run options window. For INM 6.2, a aircraft types are identified by the departure spectral class assignments its Noise-Power-Distance curves are given in Attachment 5.

4. The ability to calculate the percentage of a given boundary area covered by each contour level and output the percentage in the table produced by the Output // Contour Area and Population function has been added to the program. The percentage values replace the contour area values previously displayed in the Acres column. The boundary area is defined by the boundary.txt file which must be located in the Ambient Noise Directory defined in Setup // File Locations.

5. The heading of the last column of the pop_conr.dbf file has been changed from ACRES to PCT_BOUND. The field specifications for this column have been changed from 1 to 2 decimal places.

6. The minimum value for the "Do Percent of Time (hr)" field in the Run // Grid Setup window has been changed from 0.1 to 0.01.

7. The Setup // File Locations window has been changed to allow for the specification of the name and location of a Boundary file. The Boundary File is used to calculate the percentage of a user-specified boundary area covered by individual noise contours as described above.

8. The minimums for the max and min cutoff values in Run // Run Options have been changed to –999.9 to accommodate the new TAUD and DDOSE metrics.

9. An "Optional Export File Name Prefix" field has been added to the File // Export as Shapefile window. This field allows the specification of a file name prefix that gets added to the numerous output files created when exporting shapefiles. The prefix will make it easier to distinguish between different sets of shapefiles.

10. The File // Export as Shapefile function has been changed to produce a modified Flight-Tracks.dbf output file. The new Flight-Tracks.dbf file contains operation type, runway ID, and track ID instead of the previous track ID only. This change enhances the ability to filter track data viewed outside of the INM using ESRI shapefiles.

11. The File // Export as Shapefile function has been changed to export more detailed population data. Previously, the Population-Points shapefiles *(*.shp, *.shx,* and *.dbf)* contained Multipoint object types whose feature names were consistent with the population groupings in the "Output\\Output Graphics" Census Display control dialog box (i.e. POP <= 300 for the population points that were less than or equal to 300). The new Population-Points shapefiles contain Point type objects whose feature names contain the exact population values at the individual points (i.e. POP_37 indicates a population of 37 at the given point). The new Population-Points shapefiles will be much larger than those generated by INM version 6.1 because each individual population point is now considered an attribute.

12. The File // Export as Shapefile function has been changed to export noise data associated with standard grid points, detailed grid points, and location points. Previously, the Grid-Points and Location-Points shapefiles (*.shp, *.shx, and *.dbf) contained Multipoint object types whose feature names were consistent with the applicable grid names or location point names (i.e. Grid_S01 for all of the grid points in the standard grid named S01). The new shapefiles contain Point type objects whose feature names contain the noise metric values at the individual points (i.e. Grid_S01_67.8DNL). The new shapefiles will be much larger that those generated by INM version 6.1 because each individual grid or location point is now considered an attribute.

13. The File // Export As MIF/MID function was added to INM, as explained in the "New MapInfo Interchange File Export Function" section above.

14. The ability to filter radar tracks by runway end when creating INM tracks from radar data in the Input Graphics window has been added to the program.

15. A "Previous Zoom" button has been added to the Input and Output Graphics windows. The list of items in the View menu as well as the Input and Output Graphics buttons have been re-ordered to enhance consistency.

16. The default value of the "Refinement" contouring parameter in the Run // Run Options window has been changed from 6 to 8. The default value of the "Tolerance" contouring parameter has been changed from 1.00 to 0.25. These new default values will produce a higher-resolution contour grid and more accurate contours as compared to the previous default values.

17. A "000_None" spectral class has been added to the Acft // Noise Identifiers window. This change allows aircraft to have no Approach or Departure spectral classes as is the case with the new Maule aircraft added to INM version 6.2. Previously aircraft with no spectral class identified for a given category had their spectral class automatically assigned to a default value. A noise identifier must have a spectral class other than the "000_None" spectral class assigned for at least one of the three spectral class categories.

18. New terrain options have been added to the Run // Run Options window. When the Do Terrain box is checked, a new drop-down list appears listing three terrain data formats. A new Do Line-of-Sight Blockage box also appears.

19. The "Do Terrain" label has been changed to "Terrain Type" in the "CASE RUN OPTIONS" section of the Case Echo Report. The possible values have also changed to represent the new terrain options available in the Run // Run Options window.

20. The values of the "Do Terrain" item in the flight.txt file generated by the Output // Flight Path Report function have been changed to match the values saved in the "DO_TERRAIN" column of the case.dbf file. The possible values in the "DO_TERRAIN" column of case.dbf have been expanded to represent the new terrain options available in the Run // Run Options window.

21. The Standard Grids window generated by the Output // Standard Grids function has been modified. The USER column header has been changed to METRIC to match the column header in the grid_std.dbf file.

22. New terrain data checking has been added to Run // Run Start. When a Case containing a contour grid is run with the Do Terrain box checked and the Do Line-of-Sight Blockage box not checked in the Run // Run Options window, the INM will determine whether there is enough terrain data available in the Terrain Files directory to cover the entire contour grid. If there is not sufficient terrain data available a terrain_error.cad file showing the contour grid boundary and the boundaries of each individual terrain file is written to the Case directory. When any case is run with both the Do Terrain box and the Do Line-of-Sight Blockage box checked in the Run // Run Options window, the INM will determine whether there is enough terrain data available in the Terrain Files directory to cover a rectangle encompassing the contour grid (if applicable for the given Case), all of the applicable location points, population points, standard grid points, and/or detailed grid points, and the extent of all of the calculated flight paths. If there is not sufficient terrain data available a terrain_error.cad file showing the required Line-of-Sight Blockage terrain rectangle and the boundaries of each individual terrain file is written to the Case directory. The terrain_error.cad file can be viewed in the Tracks // Input Graphics window, with red indicating areas of missing terrain data.

23. The ability to import two new terrain data formats, Digital Elevation Model (DEM) data and National Elevation Dataset (NED) GridFloat data, has been added to the Terrain Processor under File // Import Data into Study // Terrain Files.

24. The Terrain Processor under File // Import Data into Study // Terrain Files has been modified to add additional terrain elevation grid points to the imported terrain data around the outside edge of the terrain contour rectangle. The additional points are only applied to the data imported for terrain contour viewing in the Output // Output Graphics window and have no impact on noise calculations involving terrain data. These additional points help NMPlot to close each of the terrain contours and in some cases help to more sharply define the edge of the terrain contour rectangle when the terrain contours are viewed in Output // Output Graphics.

25. The name and format of the binary file used to save imported terrain contours have been changed. The file name has been changed from _terrain.bin to _terrain62.bin. The new file format allows larger amounts of terrain contour data to be viewed in the Output // Output Graphics window. When an older Study is opened in INM 6.2, the INM will automatically move any terrain contour data in the old _terrain.bin file to the new _terrain62.bin file. The old _terrain.bin file will be automatically deleted.

B.7. Reported Problems Fixed

1. Removed duplicate thrust coefficient ID's and data for the GIIB and GIV aircraft in the INM standard database that were introduced to the database in INM 6.0c.

2. Fixed a problem with the View // Fonts function and label printing from Output Graphics. Previously Output Graphics labels for Location Points, etc. would print out in a font so small they were difficult to see. View // Fonts previously would not allow the font properties to be changed in the Output Graphics window.

3. Fixed the Radar CSV file import function to accept "0000" as a beacon code. In the User's Guide users are instructed to use "0000" as a default beacon code if they do not have an actual beacon code, and the function previously would not accept "0000".

4. Fixed a minor problem when using terrain data. Previously, for grid locations with altitude lower than airport field elevation (AFE), positive elevation angles (β) were utilized in the calculation of the lateral attenuation adjustment (LAADJ). This resulted in the calculation air-to-ground attenuation for these receivers. Currently, no air-to-ground attenuation is calculated for any receiver location when the aircraft is on the ground. This change only affects receivers below AFE when Terrain is turned on.

5. Fixed a problem with Acft // Fixed-Point Profiles window. Previously, the operational mode combo box was disabled after profiles were initially created and the INM was closed, preventing users from changing the operational modes used by the profiles.

6. Fixed a problem with the Terrain Processor when importing terrain data inside a terrain contour rectangle that has one of its corner points at the X,Y coordinates (0,0). Previously such a terrain contour rectangle would cause NMPlot to crash when generating terrain contours. Now the INM will detect if a user-specified terrain contour rectangle has a corner at the point (0,0) and will move that corner away from (0,0) to avoid causing problems for NMPlot.

7. Fixed a problem with an error message that alerts users at run time when there is only one NPD curve for a given metric and operation type. Previously this message incorrectly reported the metric identifier for the single NPD curve.

Attachment 1 – Updates to INM Noise/Performance Database

The core of the INM database consists of noise and performance under certain reference conditions. Aircraft performance profiles represent a full power takeoff for a procedure labeled as STANDARD. For the majority of the commercial transport aircraft, these procedures begin pitch over/acceleration at 1000 feet. Power cutback occurs either during or at the end of the acceleration. For many aircraft, this resembles what was once designated as ICAO B though the INM user will note some variation among aircraft. These procedures were developed for different takeoff weights that were related to the operating range of the aircraft. In developing these weights, manufactures make assumption about load factor and assumed pounds per passenger. Many aircraft developed in the late 1980's for the original INM database assumed a 60% load factor at 200 pounds per passenger and no excess cargo. Recent survey data has been shown to support higher weights per trip length and new aircraft added since 1995 have developed weight-to-stage length assumptions based on "rules" that lead to higher weights. Aircraft have also grown in maximum allowable takeoff weight since the late 1980's. A new aircraft developed today, using the old 60% load factor, would still result in a "heavier" aircraft due to this increase in maximum certification takeoff weight. This release of INM updates 5 aircraft previously developed for INM using a consistent set of takeoff procedure and weight-to-stage length rules. These weights and procedures are more consistent with current submissions, and it is anticipated that FAA in cooperation with NASA and Eurocontrol will continue to sponsor research and development that harmonizes assumptions across all of the aviation industry.

INM Standard Procedures

INM Standard procedures are provided for maximum takeoff power and maximum climb power conditions. Recent research has developed 3 types of procedures for INM. These include a standard procedure which performs engine cutback at 1000 feet Above Field Elevation (AFE) and ICAO A and ICAO B procedures. These procedures differ in the way flap retraction and power cutback occur.

Present data development of INM will continue to provide three types of procedures. However, recent surveys do not show much use of what are called "ICAO B like" procedures. ICAO A and ICAO B procedures were last defined in *Amendment 10 of the Fourth Addition of the ICAO Procedures for Air Navigation Services, Aircraft Operations, Volume 1 Flight Procedures*. This document is commonly referred to as Volume 1 of ICAO PANS-OPS. Amendment 11 of this document was published on January 11, 2001 and provided new guidance on noise abatement procedures. This document provided procedure guidance rather than specific procedures and the specific ICAO A and ICAO B procedure definitions *were not retained*. ICAO A and ICAO B procedures may still be developed under Amendment 11 guidance but the specific procedure definition is not listed in the ICAO Pans OPS. INM provides for the use of these for historical comparisons and will retain them as standard data depending on the needs of the user community.

Survey data of 747400 and 777200ER weight-to-trip length ratios demonstrates many operations in excess of the 4500 nautical mile upper limit given in previous versions of INM. INM 6.2 adds two new ranges for the distances of 4500-5500 and 5500-6500 nautical miles in order to provide more weights for these longer ranges. Some aircraft such as the 747-400 and 777-200ER, also include a 9th weight for the maximum certificated takeoff weight. For new submissions to INM, the last trip length weight will be the maximum certificated weight of the aircraft. Users should select takeoff weights based on the best data available. In the absence of such data, users may select weights based on trip length according to the rules given below.

Takeoff weight for trip length stages:

Stage No.	1	2	3	4	5	6	7	8	9
Trip Length Range (nm X 1000)	0-.5	.5-1	1.0-1.5	1.5-2.5	2.5-3.5	3.5-4.5	4.5-5.5	5.5-6.5	>6.5
Representative Range	350	850	1350	2200	3200	4200	5200	6200	
Weight (lb X 1000)	___	___	___	___	___	___	___	___	___

In developing takeoff weights for stage lengths, the following guidance has been established to provide common mission planning rules for determining default weights. These "rules" have been shown to correlate with survey data. Airlines do not always purchase aircraft at their maximum certificated weight. INM aircraft are developed based on the maximum certificated weight, with the weight provided as a lower bound on the climb performance of the aircraft.

Parameter	Planning Rule
Representative Trip Length	Min Range + 0.70*(Max Range – Min Range)
Load Factor	65% Total Payload of the Maximum Certificated weight sold to airlines.
Fuel Load	Fuel Required for Representative Trip Length + the average of ATA Domestic and International Reserves
	As an example, typical domestics reserves include 5% contingency fuel, 200 nm alternate landing with 30 minutes of holding.
Cargo	No additional cargo over and above the assumed payload percentage

INM STANDARD Procedure:

1) Takeoff at Full power
2) Cutback to climb power around 1000 feet AFE and pitch-over to accelerate
3) Accelerate to clean configuration
4) Climb to 3000 feet AFE
5) Accelerate to 250 knots
6) Continued climb to 10000 feet AFE

INM ICAO A Procedure:

1) Takeoff at Full Power
2) Climb to 1500 feet AFE at full power holding flaps
3) Cutback to Climb Power at 1500 feet
4) Climb to 3000 feet AFE at climb power holding flaps
5) Accelerate to clean configuration
6) Accelerate to 250 knots
7) Continued climb to 10000 feet AFE

INM ICAO B Procedure:

1) Takeoff at Full Power
2) Climb to 1000 feet and pitch-over to accelerate
3) At full power, accelerate to clean configuration
4) Cutback to climb power
5) Climb to 3000 feet AFE
6) Accelerate to 250 knots
7) Continued climb to 10000 feet AFE

Attachment 2 – Ambient Data Input Files

Supplemental metrics such as TAUD require input data files that contain estimates of ambient sound levels. There are two types of data that may be collected. The first contains representative A-weighted sound levels assigned to a regularly spaced grid and is referred to as the *ambient grid* file. The other contains representative 1/3-octave band data that is also assigned to a regularly spaced grid through an indexing convention described below. This is referred to as the *ambient spectral map* file. The location and actual filename of the ambient grid file must be specified using the *Setup // File Locations* dialog window (see *Ambient Noise File* box). The ambient spectral map file must be named *ambi_map.txt* and reside in the same directory as the ambient grid file. Example ambient grid ("ambient.txt") and ambient spectral map files are included as a part of this attachment.

NOTE: The Ambient Grid File described below was first introduced in INM 6.0 and is documented on page 10-5 of the INM 6.0 Users Guide. The format given in this Attachment now supports ambient data to one decimal place (i.e., 3 total digits) whereas the original format specified only integer values (2 total digits). Files developed in the old format are still supported for backward compatibility for TALA. However, for TAUD, INM automatically converts data files to the new 3-digit format. 2-digit data are archived in a file called ambient_backup.txt and the new 3-digit data are written to Ambient.txt

Ambient Grid File

The purpose of the ambient grid file is to assign a number, representing the A-weighted ambient sound level, to study area grid points. This file is a space-delimited, ASCII text file with format and use illustrated with an example file at the end of this Attachment. The first five rows contain header information that gives the specific dimensions of the grid which is referenced to a latitude/longitude coordinate system. The first two rows, *ncols* and *nrows*, give the number of columns and rows of the regular grid. The third and fourth rows give the Lower Left (ll) or southwest corner of the grid in terms of latitude/longitude in decimal degrees. Row 3 contains the field id "xllcorner" followed by a real number specifying the longitude (x-coordinate of grid) in decimal degrees. Row 4 contains the field id "yllcorner" followed by a real number specifying the latitude (y-coordinate of grid) in decimal degrees. The fifth row contains the field id "cellsize" followed by a real number specifying the spacing between both latitude and longitude points in decimal degrees. The final grid in this example will contain a 15 column by 12 row array of points, evenly spaced 0.1 decimal degrees apart referenced to a lower-left (southwest) corner of -114.03464052 longitude and 35.61089089 latitude.

The sixth row contains the text "NODATA_value" followed by an integer. This value is used to indicate that no ambient grid data are available for one or more locations within the grid. When computing TAUD for locations specified as having no data, by default the INM assigns the ISO threshold of human hearing spectral data to those locations.

Lines 7 through X+7 each contain Y three-digit integers. The integers represent A-weighted sound levels and are stored as ten times the value they represent (i.e., '347' represents 34.7 dB).

Ambient Spectral Map

The ambient spectral map file is a comma-delimited, ASCII text file which assigns spectral data to the grid points contained in the ambient grid file above. The first row contains an integer specifying the number of data rows which follow. Each row contains the following information: (1) first field: index of spectrum, for informational purposes only and *not* used at this time; (2) second field: spectrum name/site name, for informational purposes only; (3) third field: A-weighted energy sum of spectrum. This value should have a corresponding match in the ambient grid file above; (4) fields four through twenty-seven: sound pressure levels for one-third octave bands 17 (50 Hz) through 40 (10,000 Hz). Note that field 3 above is the value which is indexed with the ambient grid file for specifying grid-based ambient spectra. The index convention maps a field 3 value of 34.7 to all values of 347 in the ambient grid file. It is useful for documentation purposes for this value to be equivalent to the A-weighted sum of the spectrum, however this is not required and the convention may not hold for the rare case when different spectra have identical A-weighted values. Regardless of convention, the values of column 3 must be unique across all rows. To assist the user, the INM calculates the A-weighted sum of each spectrum and compares it to the reported value (#3 above). If the two values are not equivalent to within +/- 0.1 dB, a message is printed to the *ambient_warning.txt* file, and the program continues, using the specified spectral data.

Sample Ambient Grid Text File – (*Ambient.txt*)

ncols 15
nrows 12
xllcorner -114.03464052
yllcorner 35.61089089
cellsize 0.1
NODATA_value -99
347 347 347 347 347 347 347 347 347 347 347 347 347 347 347
347 347 347 347 347 347 347 347 215 347 347 347 347 347 347
347 347 347 347 347 347 347 347 215 215 215 347 347 347 347
347 347 347 347 347 345 345 345 215 347 347 347 347 215 215
347 345 345 345 345 345 345 345 347 347 347 347 347 215 215
347 345 345 345 345 345 345 347 347 347 347 347 347 215 215
347 347 347 345 347 347 347 347 347 347 347 347 347 347 347
347 347 347 347 347 347 347 347 347 347 347 347 347 347 347
347 347 347 347 347 347 347 347 347 347 228 347 347 347 347
347 347 214 347 347 205 205 205 347 347 347 228 228 228 228
347 347 214 214 347 205 205 205 347 347 347 228 228 228 228
347 347 347 347 205 205 205 347 347 347 347 228 228 228 228
Sample Ambient Spectral Map Text File – (*ambi_map.txt*)

6
1,3A-1,34.7,45,39.7,35.7,32.7,30.9,30.8,30.8,29.9,29.6,29.6,29.2,28.6,27.8,27.2,26.4,24.6,21.9,19,14.5,9.9,8,7.2,14.8,23.6
2,3A-2,34.5,45,39.5,35.2,32.1,30.3,30.3,30.4,29.5,29.5,29.5,29.2,28.7,28,27.2,26.1,24,21,17.5,12.8,8.7,8,9.4,14.8,23.6
3,3B-2,22.8,44.9,39.2,34.3,30.5,27.7,25.5,23.6,22.2,21,20,18.5,17.5,16,15.2,14.7,13.6,12.3,10.6,8.5,6.7,7,7.2,14.8,23.6
4,3B-2,21.4,44.9,39.1,34.1,29.9,27,24.8,22.7,21.4,20.1,19,17.6,16.5,15.2,14.4,13.9,12.9,11.6,10,8.1,4.6,4.4,7.2,14.8,23.6
5,3D-1,21.5,44.9,39.1,33.9,29.2,25.8,22.6,19.6,17.7,16,14.8,14.3,14.2,14.6,15,15.4,14.9,14.1,12.6,10.1,7.4,4.4,7.2,14.8,23.6
6,3D-2,20.5,44.9,39.1,33.9,29.3,25.9,22.6,19.6,17.6,16,14.7,14.2,14,14.3,14.6,14.9,14.4,13.3,11.6,9.1,4.6,4.4,7.2,14.8,23.6

Attachment 3 – Calculating Audibility

Introduction

Audibility is defined as the ability for an attentive listener to hear aircraft noise. Detectability is based on signal detection theory[63,64], and depends on both the actual aircraft sound level ("signal") and the ambient sound level (background or "noise"). As such, audibility is based on many factors including the listening environment one is in. Conversely, detectability is a theoretical formulation based on a significant body of research. For the purposes of INM modeling the terms "audibility" and "detectability" are used interchangeably. The detectability level (d') calculated in INM is based on the signal-to-noise ratio within one-third octave-band spectra for both the signal and noise, using a 10log(d') value of 7 dB.

There are three parts to the calculation of audibility in INM: (1) Calculate the detectability level ($D'L_{band}$) for each one-third octave band of the signal for a single contributing flight path segment; (2) Calculate the detectability level ($D'L_{total}$) for the overall signal for a single contributing flight path segment; and (3) Calculate absolute or percentage of time a signal is audible (detectable by a human) for a flight path (*TAud* or *%TAud*).

Definitions

$L_{signal,band}$	sound level of the signal (aircraft) for a particular frequency band
$L_{noise,band}$	sound level of the background noise (ambient) for a particular frequency band
η_{band}	efficiency of the detector (a scalar value known for each frequency band)
Bandwidth	1/3-octave bandwidth
$D'L_{band}$	detectability level for a particular frequency band
$D'L_{total}$	total detectability level
d'_{band}	detectability for a particular frequency band
d'_{total}	sum of squares of detectability over all frequency bands
Taud	absolute amount of time a signal is audible by humans
%Taud	percentage of a time period that a signal is audible

Note that values of $L_{signal,band}$ and $L_{noise,band}$ are calculated for each segment-receiver pair and then the total audibility for a flight track is summed from the individual segments.

Part I of Calculations:

Calculate the detectability level for each one-third octave frequency band, then determine if the signal for that frequency band is detectable.

The theory of detectability level is based on the following equation:

$$D'L_{band} = 10\log\left[\eta_{band}\sqrt{bandwidth}\left(\frac{signal}{noise}\right)\right] \qquad \text{C-1}$$

The following one-third octave band filter characteristics are used in the calculation of detectability:

Table B-1. One-Third Octave Band Characteristics

ANSI Band #	Nominal Center Frequency (Hz)	Bandwidth (Hz)	$10\log[\eta_{band}]$
17	50	11	-6.96
18	63	15	-6.26
19	80	19	-5.56
20	100	22	-5.06
21	125	28	-4.66
22	160	40	-4.36
23	200	44	-4.16
24	250	56	-3.96
25	315	75	-3.76
26	400	95	-3.56
27	500	110	-3.56
28	630	150	-3.56
29	800	190	-3.56
30	1000	220	-3.56
31	1250	280	-3.76
32	1600	400	-3.96
33	2000	440	-4.16
34	2500	560	-4.36
35	3150	750	-4.56
36	4000	950	-4.96
37	5000	1100	-5.36
38	6300	1500	-5.76
39	8000	1900	-6.26
40	10000	2200	-6.86

1) Calculate the detectability level for each 1/3-octave frequency band

$$D'L_{band} = (L_{signal,band} - L_{noise,band}) + \{10\log[\eta_{band}] + 0.5 \times 10\log[bandwidth]\} \qquad \text{C-2}$$

Where $10\log[\eta_{band}]$ is given in the above table.

2) Determine if the signal for that frequency band is detectable

   If $D'L_{band} \geq 7$ the signal is flagged as **detectable** for that frequency band

Part II of calculations

Determine if the overall signal is detectable.

1) Calculate the detectability for each one-third octave frequency band using the band detectability levels from Part I

$$d'_{band} = 10^{\frac{D'L_{band}}{10}} \qquad \text{C-3}$$

2) Calculate the square root of the sum of squares of detectability over all frequency bands

$$d'_{total} = \sqrt{\left[\sum_{band=17}^{40}(d'_{band})^2\right]} \qquad \text{C-4}$$

3) Calculate the total detectability level

$$D'L_{total} = 10\log[d'_{total}] \qquad \text{C-5}$$

4) Determine if the overall signal is detectable

If $D'L_{total} \geq 7$ the overall signal is flagged as **detectable**

Part III of calculations:

Calculate the absolute or percentage of time a signal is audible by a human; the time for a single contributing flight path segment is first calculated, then the absolute or percent time is calculated for an overall event or larger period of time (multiple flights for an average day or other time period)

1) Calculate the time audible (in seconds) for a single flight path
   seglength length (in feet) of contributing flight path segment
   segspeed average speed (in feet/second) during contributing flight path segment
   segtime time passed during contributing flight path segment
   totaltime total time of flight for a single event

For each segment, calculate time it takes aircraft to travel through flight path segment

   segtime = (seglength/segspeed) x # of Operations

If segment is flagged as detectable, then

   TAud = TAud + segtime

Then when ***segtime*** is totaled for all segments, a 24 hour percent Time Audible will be:

   %TAud = TAud/(24 hours)

2) Calculate the time audible (in minutes) for a time period

   TAud = TAud/(60seconds/minute)

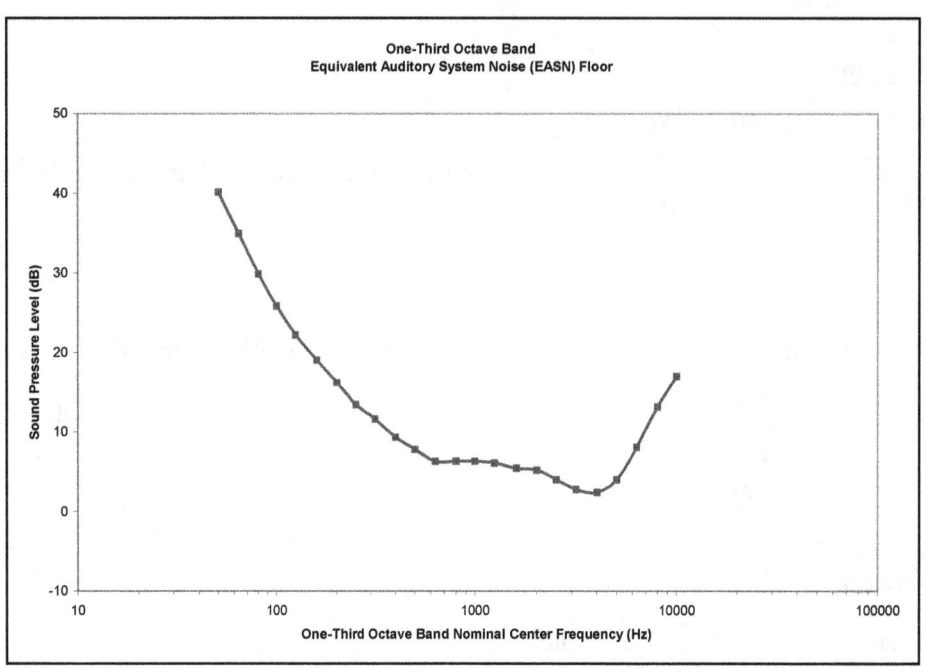

Figure B-2. One-Third Octave Band Equivalent Auditory System Noise (EASN) Floor

Data are presented spanning 20 to 20,000 Hertz. The solid portion of the curve (50 to 10,000 Hertz) represents the frequency range utilized by the INM.

Table B-2. Equivalent Auditory System Noise (EASN)

One-Third Octave Band Nominal Center Frequency (Hz)	Equivalent Auditory System Noise (EASN) (dB)
50	40 2
63	35.0
80	29.8
100	25.8
125	22 2
160	19.0
200	16 2
250	13.4
315	11.6
400	9.3
500	7.8
630	6.3
800	6.3
1000	6.3
1250	6.1
1600	5.4
2000	5.2
2500	4.0
3150	2.8
4000	2.4
5000	4.0
6300	8.1
8000	13 1
10000	17.0

Attachment 4 - INM Technical Manual Update Addendum

INM Version 6.1 modified the lateral attenuation algorithms contained in the model to better correlate modeling predictions with research undertaken recently in both the U.S. and internationally. The lateral attenuation algorithms utilized in the INM are based on SAE-AIR-1751 which specifies an algorithm with two primary components: (1) Overground Attenuation [G(ℓ)]; and (2) Long-Range Air-to-Ground Attenuation [Λ(β)]. INM 6.1 includes changes only to Λ(β) (Long-Rang Attenuation) for Wing-Mounted and propeller aircraft.

Figure B-2 below depicts the Long-Range Air-to-Ground Attenuation algorithm. The solid line (designated as "SAE-AIR-1751 (1981, reaffirmed 1991)") represents this equation as specified in SAE-AIR-1845 and used for modeling all aircraft in INM prior to Version 6.1. This curve is identical to Figure 3 in SAE-AIR-1751. The dashed line (designated as "INM Version 6.1") represents the Long-Range Air-to-Ground Attenuation used in Version 6.1 for all aircraft *except* jet aircraft with tail-mounted engines. INM Version 6.1 still uses the existing SAE-AIR-1751, represented by the solid line for jet aircraft with tail-mounted engines.

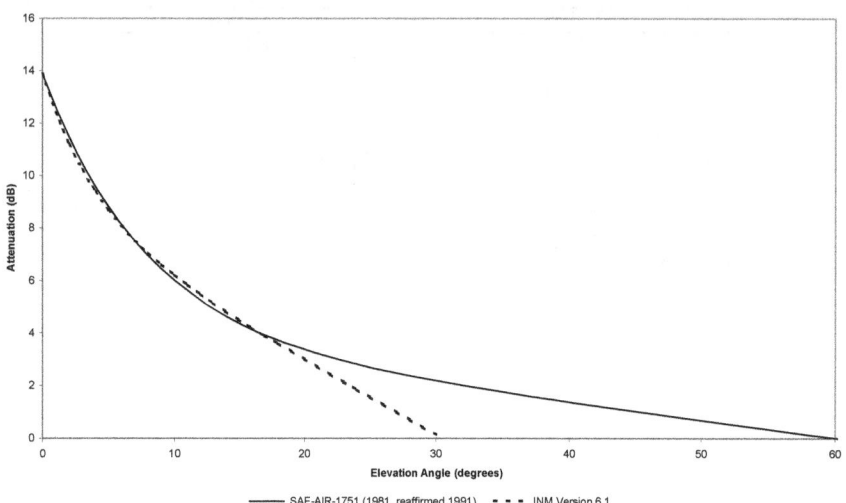

Figure B-3. Long-Range Air-to-Ground Attenuation $\Lambda(\beta)$

The Version 6.1 Long-Range Air-to-Ground attenuation algorithm may be represented by the following equation:

$$\Lambda(\beta) = a_1 + a_2\beta + a_3^{(a_4\beta)}$$

where: β = elevation angle, $0° \leq \beta \leq 30°$
a_1 = 8.66213
a_2 = -0.28436
a_3 = 5.21353
a_4 = -0.25718

Attachment 5 – Excess Lateral Attenuation and Aircraft Spectral Class Assignments

SAE-AIR-1751 and its draft update provide framework for determining excess lateral attenuation for fixed-wing aircraft. This excess attenuation has been observed from multiple field tests that have been conducted and reported to SAE. In general, these tests focus on commercial jet aircraft. Excess attenuation for military aircraft is determined by equations given in the USAF NoiseMap program. Table B-3 summarizes the excess attenuation effects for INM. Note that the user may disable ground-to-ground attenuation for aircraft designated as props. While SAE-AIR-1751 is not directly applicable to helicopters, Table 1 summarizes the application to helicopters in INM. As noted in the table, the ability to disable ground-to-ground attenuation also applies to helicopters. The "1751 Interim Update" identified in Table 1 refers to the "New Lateral Attenuation Function" introduced in INM Version 6.1.

Table B-3. INM 6.2 Lateral Attenuation Algorithm Update

	"All-Soft-Ground"		"No-Prop-Attenuation"	
	Air-to Ground	Ground-to-Ground	Air-to Ground	Ground-to-Ground
Wing-Mount Jets	1751 Interim Update	1751	1751 Interim Update	1751
Tail-Mount Jets	1751	1751	1751	1751
Props	1751 Interim Update	1751	1751 Interim Update	*none*
Helis	1751 Interim Update	1751	1751 Interim Update	*none*
Military	NoiseMap		NoiseMap	

Users creating user-defined aircraft should be aware of the relationship between NPD curve, spectral class assignment and the excess attenuation modeled in INM. Table B-4 presents the classification, by aircraft type, for the assignments of each INM spectral class.

Table B-4. Spectral Class Assignments by Aircraft Type

	Spectral Classes[5]		
	Departure	**Approach**	**Flyover / Afterburner**
Wing-Mount Jets	101-108	202-209	N/A
Tail-Mount Jets	113, 132-134	201, 216	N/A
Props	109-112	210-215	112, 213, 801-806
Helis	114-120	217-222	301-307
Military	121-131	223-234	121, 125-128, 131

---

[5] Spectral classes 801 through 806 contain data collected during measurements of aircraft noise in the National Parks.

Appendix C: Summary of Commercial Jet Overflights in GCNP, August 31, 2003

Introduction

In support of the joint US Department of Transportation (DOT), Federal Aviation Administration (FAA) and US DOI National Park Service (NPS) Alternative Dispute Resolution (ADR) process related to aircraft overflights of Grand Canyon National Park (GCNP), noise modeling sensitivity runs have been undertaken using both the FAA's Integrated Noise Model (INM) and the NPS' NoiseMap Simulation Model (NMSim). One of the modeled scenarios includes the operations on August 31, 2003, considered to be the average day of a peak month in terms of GCNP tour operations for 2003. This appendix presents a graphical summary of the ground tracks of the high-altitude jet overflights of GCNP for the top seven airports in terms of operations. The data source is the Enhanced Traffic Management System (ETMS) housed at the US DOT Volpe Center. Overflights captured for this analysis include all flights whose ground tracks intersect the GCNP INM analysis window plus a 20 nautical mile buffer.

Initially five airports representing approximately 50% of the flights over the park were selected. Two more were added after visual inspection of the overflight data. The seven airports are: Chicago O'Hare International Airport (ORD), Denver International Airport (DEN), John F. Kennedy International Airport in New York (JFK), McCarran International Airport (LAS), Los Angeles International Airport (LAX), Phoenix Sky Harbor International Airport (PHX) and Salt Lake City International Airport (SLC). The following figures present both departure operations (in blue) and arrival operations (in red) for specific scenarios. Note that in total there were 1,371 overflights of GCNP identified in the ETMS data. Because some flights may have and origin *and* a destination at the seven highlighted airports, the individual flights do not total to this value.

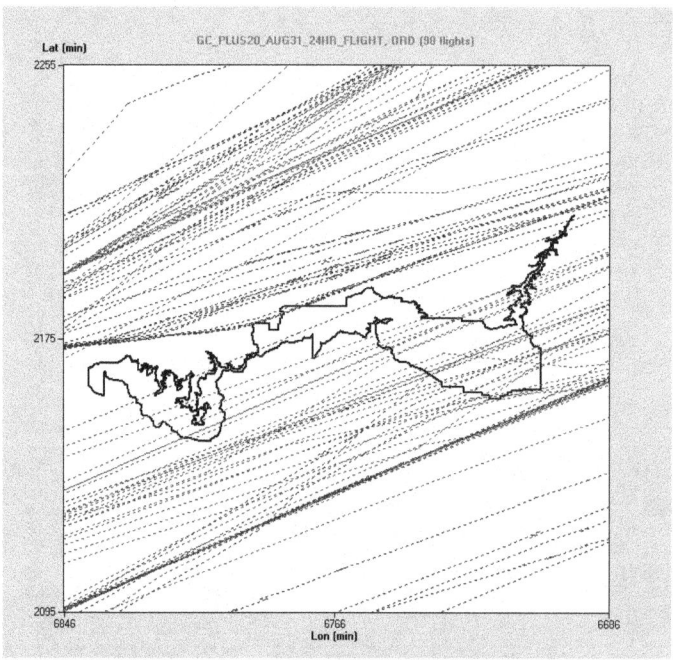

Figure C-1. ORD Flights over GCNP

(98 flights, 7% total daily departures; 6% total daily arrivals)

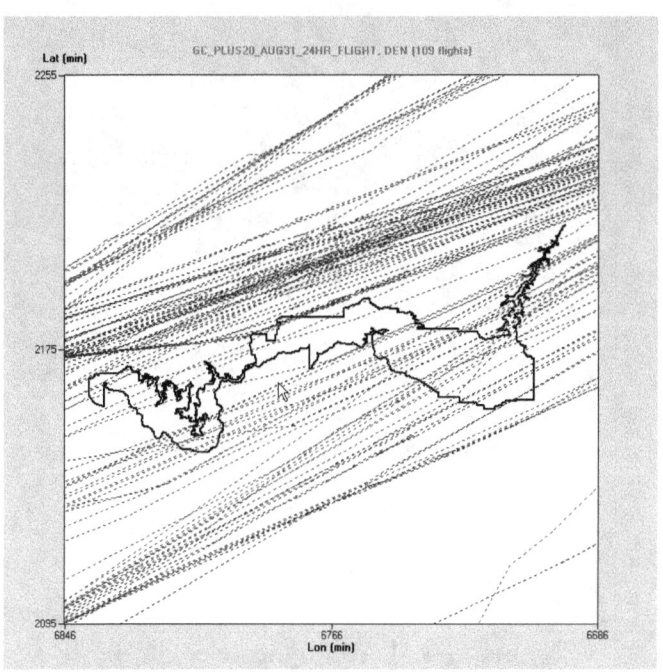

Figure C-2. DEN Flights over GCNP

(109 flights, 8% total daily departures; 7% total daily arrivals)

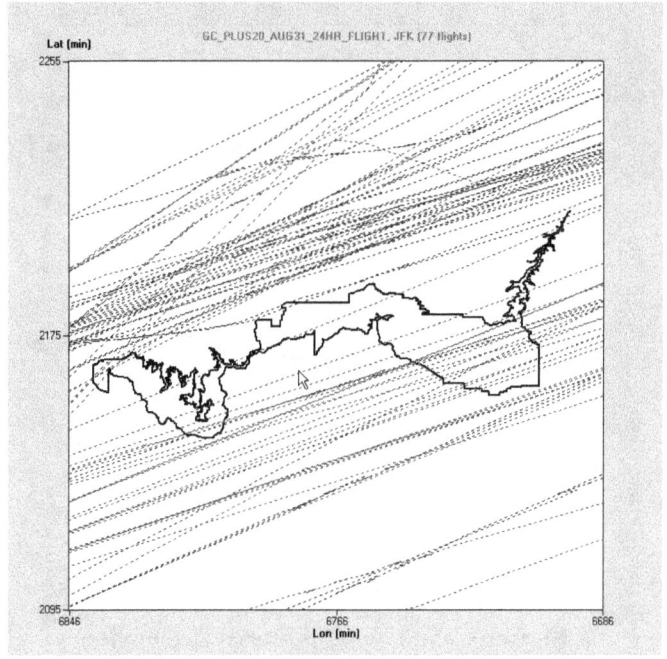

Figure C-3. JFK Flights over GCNP

(77 flights, 5% total daily departures; 5% total daily arrivals)

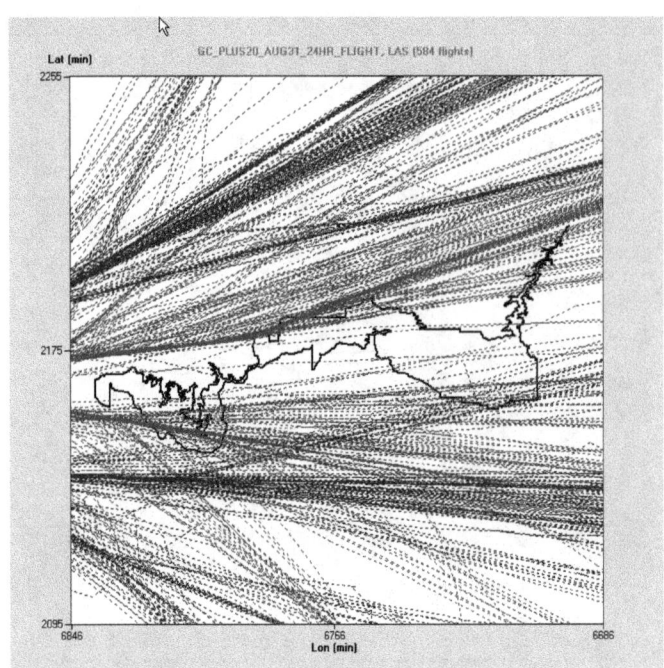

Figure C-4. LAS Flights over GCNP

(584 flights, 39% total daily departures; 39% total daily arrivals)

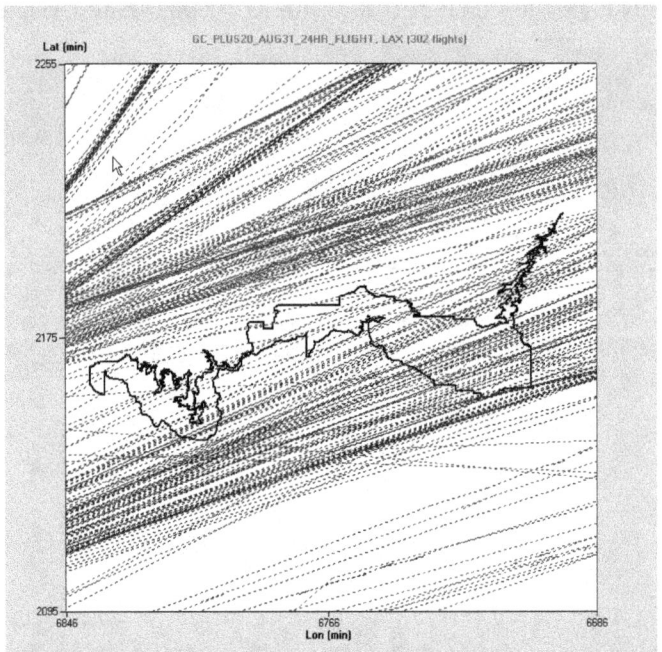

Figure C-5. LAX Flights over GCNP

(302 flights, 19% total daily departures; 21% total daily arrivals)

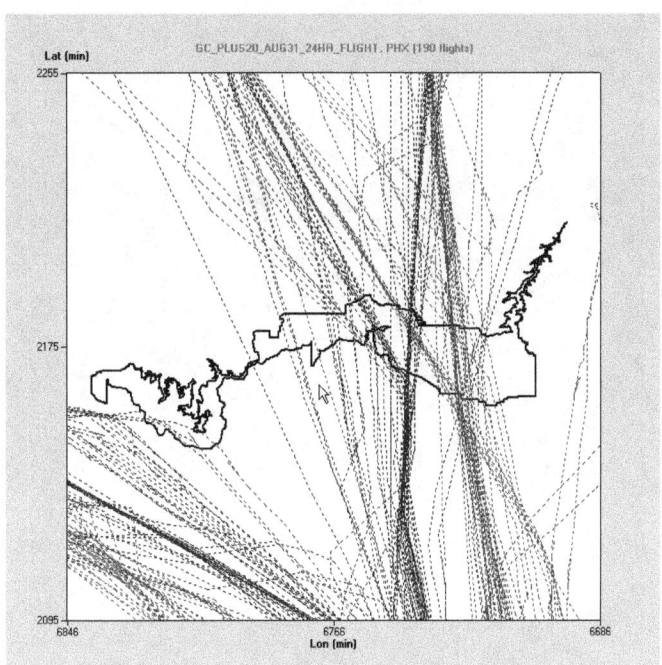

Figure C-6. PHX Flights over GCNP

(190 flights, 12% total daily departures; 14% total daily arrivals)

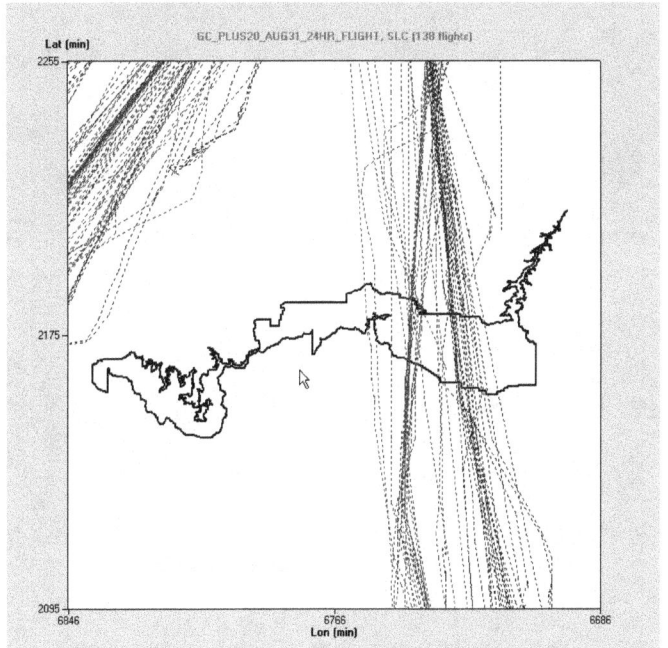

Figure C-7. SLC Flights over GCNP

(138 flights, 10% total daily departures; 9% total daily arrivals)

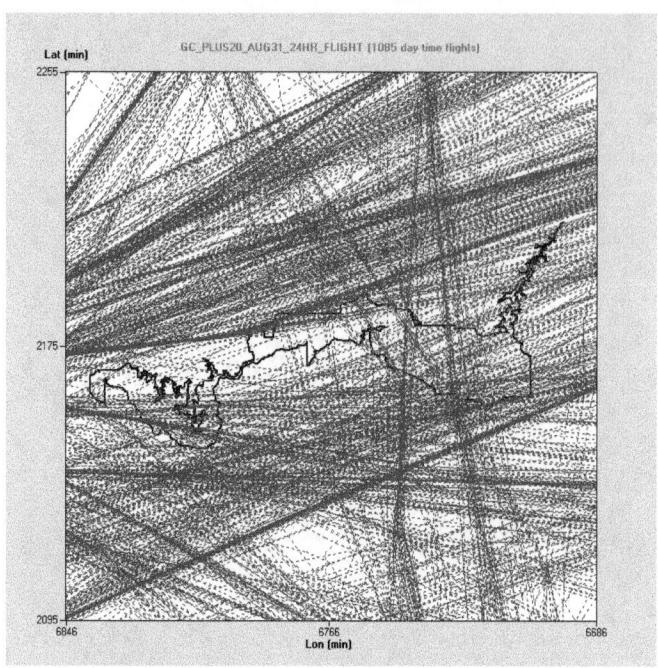

Figure C-8. All Flights over GCNP
Daytime (1085 flights)

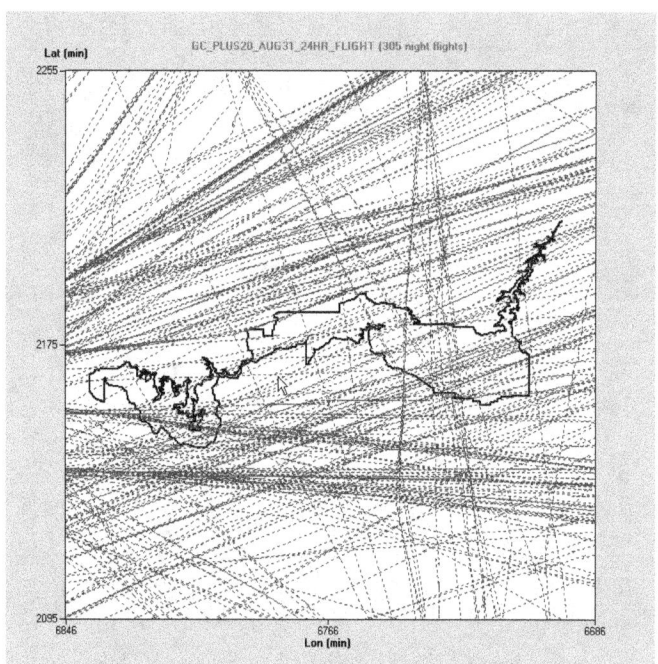

Figure C-9. All Flights over GCNP
Nighttime (305 flights)

As noted above, ETMS data were used as the primary source of overflight operational data for this study. PDARS data, an alternative source of similar data, was identified during the course of the current study. While the two data sets were considered to be somewhat similar, the project team was notified that PDARS may have better coverage for military aircraft. Accordingly, ETMS and PDARS data sets were both obtained and compared for the August 31, 2003 test day. Table C-1 below presents a summary of the differences between the two data sets for this time period for the analysis area around GCNP. Note that because the ETMS and PDARS systems have different data fields and classification systems, not all data are directly comparable and are labeled not applicable (N/A) in Table C-1.

Table C-1. Comparison of ETMS and PDARS Data

Aircraft Segment	Counts			
	ETMS	**PDARS**	in ETMS, not in PDARS	in PDARS, not in ETMS
Total	1371	1441	95	165
Commercial	1138	1269	N/A	N/A
GA	122	163	16	57
Military	0	9	0	9

Appendix D: Theoretical GCNP Jet Audibility Assessment

Introduction

In support of the joint US Department of Transportation (DOT), Federal Aviation Administration (FAA) and US DOI National Park Service (NPS), a theoretical assessment of jet aircraft audibility has been undertaken. Specifically, the US DOT's Volpe Center Acoustics Facility, in coordination with Wyle Laboratories, has analyzed the audibility of jet aircraft maximum spectra considered to be representative of high altitude commercial jet overflights of GCNP. These spectra have been modeled as propagating through atmospheric conditions considered representative of those near GCNP and the audibility of resulting spectra have been analyzed relative to typical ambient spectral data for GCNP. This appendix summarizes the analysis and presents its conclusions.

Performance Data

The FAA's Integrated Noise Model (INM) is an internationally accepted noise prediction tool, originally designed for use in the vicinity of commercial airports. The performance equations in INM were used to determine the appropriate corrected net thrust value (F_n/δ) for six representative jet aircraft in straight, level flight, at a range of altitudes spanning 20,000 to 40,000 feet MSL, as shown in Table D-1. A primary assumption in that process was that the flaps-retracted constant drag coefficient (R) in INM, which was originally developed using lower speed, terminal-area data, is representative of higher speed conditions. This is considered to be a reasonable, first-order approximation. A similar approximation is also used in the FAA's Noise Integrated Routing System (NIRS). It is also assumed that the aircraft are flown at constant indicated airspeed, so that true airspeed increases with increasing altitude. The use of digital flight data recorder (CFDR) data assembled for the FAA's System for Assessing Aviation's Global Emissions (SAGE) project was considered to be an alternative approach and may be further investigated in the future to confirm the validity of these values and assumptions. The values in the table with an asterisk were calculated using the average of the uncorrected thrusts at the lower altitudes divided by the pressure ratio at the indicated altitude.

Table D-1. Aircraft Corrected Net Thrust (F_n/δ, pounds) as a Function of Altitude

Aircraft	Aircraft MSL Altitude (feet)				
	20,000	25,000	30,000	35,000	40,000
A320	7882.5	9761.2	12197.8	15393.8	19570.6*
EMB145	2789.6	3454.5	4316.9	5448.6*	6926.3*
MD83	8982.3	11123.2	13899.8	17541.8	22301.3*
737-300	6921.1	8570.7	10710.1	13516.4	17183.7*
737-700	8060.1	9981.3	12472.8	15740.9	20011.8*
777-200	20318.3	25161.1	31441.9	39680.2	50619.9

Based on the F_n/δ data presented in Table D-1, archival manufacturer spectral data were used to interpolate/extrapolate the appropriate spectra for each altitude and power setting. This procedure is described in more detail in Attachment 2. All spectra represent sound levels at a distance of 1000 feet from the aircraft.

Propagation of these spectra to the ground gave physically unrealistic results that the higher altitude aircraft had more received noise on the ground than the lower altitude aircraft. Although these result follow from the assumptions used in this assessment-level analysis, these assumptions will need to be re-examined prior to modeling of actual operations.

All spectra were also scaled in accordance with a National Aeronautics and Space Administration (NASA) paper by Shepherd and McAninch. That scaling process is also detailed in Attachment 2.

The aircraft spectra were then propagated from the altitudes shown in the above table to ground level, assuming both divergence and atmospheric absorption. Atmospheric absorption was computed using a layered atmosphere and representative meteorological data from a National Oceanographic and Atmospheric Administration (NOAA) balloon launch at Flagstaff, Arizona.

The audibility of each propagated aircraft spectrum was then evaluated using two GCNP ambient spectra from the GCNP MVS. One spectrum was taken from a site with relatively high low-frequency noise due to noise from nearby rapids, the other had relatively high mid-frequency noise due to wind noise in nearby trees. These spectra are conservative in the sense that they represent cases where the aircraft would be least audible. Similar to previous GCNP analyses, the calculation of audibility utilized the detectability level (d') based on the signal-to-noise ratio within one-third octave band spectra using a 10log(d') value of 7 dB.

Conclusions

For the combination of aircraft and ambient spectra, all aircraft were calculated to be audible at all altitudes (up to 40,000 ft MSL). Moreover, no 10log(d') less than 34 dB was calculated, indicating that the high altitude jet aircraft are clearly audible in the GCNP at the time of A-weighted maximum sound level. Final quantitative results will be developed after the issues cited above are addressed. The audibility result does, however, match the experience of the measurement teams in the field on July 16-19, 2004.

Attachment 1: Extension of "On the use of corrected net thrust to estimate jet noise"

Introduction

"On the use of corrected net thrust to estimate jet noise" by McAninch, Shepherd, and Rawls provides a method of modifying the Lighthill equation on jet noise to account for the Integrated Noise Model's use of net corrected thrust. A copy of the relevant equation from Appendix A of the paper follows below.

$$\langle p^2 \rangle = \left\{ \left[\frac{K_p}{r_0^2 A_j \rho_a c_a^5} \right] \left[\frac{1}{(1+R_D/F_n)^2} \right] \left[\frac{1}{\rho_a U_a A_i} \right]^4 F_n^6 \right\} (\rho_0 c_0) \tag{1}$$

Equation (1) represents a general relationship between thrust and sound pressure. We can consider the specialized case of cruise flight, where the forces acting on the aircraft balance, and simplify this general relationship.

Extension to cruise flight

For aircraft in steady cruise flight, we assume that the total thrust produced by the aircraft engines equals the total drag of the aircraft.

$$F_n = Drag = C_D q S \tag{2}$$

where F_n represents the aircraft's total net thrust, C_D represents the drag coefficient, q equals the dynamic pressure, and S represents a characteristic area (typically the wing area). Conventional aerodynamic theory defines the dynamic pressure q as:

$$q = \frac{1}{2}\rho V^2 \tag{3}$$

or, in the notation of the McAninch paper,

$$q = \frac{1}{2}\rho_a U_a^2. \tag{4}$$

The ram drag over total thrust term in equation (1) above, when combined with equation (4), becomes:

$$\frac{1}{\left(1+\frac{R_D}{F_n}\right)^2} = \frac{1}{\left(1+\frac{2\rho_a U_a^2 A_i}{C_D S \rho_a U_a^2}\right)^2} = \frac{1}{\left(1+\frac{2A_i}{C_D S}\right)^2}. \tag{5}$$

At high speeds, conventional aerodynamic theory states that the parasitic drag component of C_D has a larger magnitude than the induced drag component. If the parasitic drag component dominates C_D, then C_D does not depend on airspeed, and all the terms in equation (5) remain constant until the airspeed approaches the critical Mach number.

If we substitute equation (5) into equation (1) and drop the constant terms, we can compare the received sound pressure at two arbitrary altitudes:

$$\frac{\langle p_a^2 \rangle}{\langle p_0^2 \rangle} = \left\{ \frac{(1/\rho_a c_a^5)(1/\rho_a U_a)^4 (\rho_a U_a^2)^6}{(1/\rho_0 c_0^5)(1/\rho_0 U_0)^4 (\rho_0 U_0^2)^6} \right\} = \left\{ \frac{(\rho_a U_a^8/c_a^5)}{(\rho_0 U_0^8/c_0^5)} \right\} \tag{6}$$

We can convert this ratio of pressures to a decibel difference by multiplying the logarithm of each side by ten:

$$\text{decibel difference} = 80 \log (U_a/U_0) + 10 \log (\sigma) - 25 \log (\theta) \tag{7}$$

where σ represents the density ratio, θ represents the temperature ratio, and we can represent the speed of sound c as a function of the square root of the temperature $(c = f\{\sqrt{T}\} = k\sqrt{T})$.

If we let the aircraft maintain a constant indicated airspeed, then the true airspeed will increase as the aircraft gains altitude. The true airspeed equals the indicated airspeed divided by the square root of the density ratio at the particular altitude:

$$U_{\text{true}} = \frac{U_{\text{indicated}}}{\sqrt{\sigma}}.$$

We can apply this relationship to equation (7). Doing this further simplifies the decibel difference equation to:

$$\text{decibel difference} = -30 \log(\sigma) - 25 \log(\theta) \qquad (8)$$

Attachment 2: Correcting Low-Altitude Jet Noise to Higher Altitudes

Introduction

This brief report discusses the development of a technique for modifying aircraft noise collected during low-altitude, low-speed tests to correct for high-altitude, high-speed conditions. The report also presents some implications of using this technique. The report concludes with a comparison of those noise values extrapolated to high-altitude, high-speed conditions using the current technique with noise values extrapolated using the standard INM.

McAninch, Shepherd, and Rawls, in a paper entitled "On the use of corrected net thrust to estimate jet noise," developed a method for replacing the jet velocity term in the Lighthill jet noise equation with thrust. Once this substitution has been made, users of the McAninch model have a means of testing the effects of changing parameters on the jet component of aircraft noise. For the present study, where we seek to determine the correlation of high altitude aircraft overflights and noise impacts at long propagation distances, the McAninch model may prove useful. This usefulness results from a confluence of physical relationships: 1) jet noise contributes primarily to the low frequency content of aircraft noise, 2) low frequency noise propagates through the atmosphere with less attenuation than higher frequencies, and 3) measurements have shown that the low frequency component of an aircraft's noise spectrum determines aircraft's audibility to human observers at long distances. The ability to predict the influence of parameter changes on thrust, and therefore on jet noise, constitutes a primary contribution to the study of the noise impacts due to high altitude overflights.

Cruise Conditions

We can simplify the McAninch model by assuming the aircraft of interest fly in a cruise configuration. The primary benefit of this assumption comes from the elimination of aircraft physical characteristics from the model, so that only atmospheric characteristics remain. A copy of a brief description of this simplification accompanies this report. The following equation represents the relationship between the atmospheric parameters at the aircraft's altitude and the noise impact on an observer on the ground:

$$\Delta dB = -30\log(\sigma) - 25\log(\theta),$$
where: σ represents the density ratio, and
θ represents the temperature ratio.

In addition to the assumption that the aircraft's physical characteristics remain constant, the above equation also assumes that the pilots fly their aircraft at a constant indicted airspeed. For this case, the aircraft increases true airspeed when climbing and decreases true airspeed when descending.[6] Note that the INM uses a fixed value of the drag-over-lift coefficient 'R', and does not take into account that more thrust is required to fly faster at higher altitudes. We therefore expect that the INM will predict lower noise levels than the McAninch model.

---

[6] If the pilots do not hold a constant indicated airspeed, then they need to change thrust settings. If this happens, our ability to predict thrust, and therefore noise, vanishes.

Note that the simplified McAninch model contains no terms related to frequency content or directivity. The original Lighthill model also contains no terms of these types. These models predict total sound power, not components of that power. For this reason, we only use the McAninch model to predict the maximum sound received by the observer, not to predict a time history nor a spectral composition.

Application of the McAninch model

In the present study, we apply the McAninch model in the following way:

1) Using the INM, we calculate an L_{ASmx} value at 1000 feet MSL for each of the six aircraft types of interest in this study. The INM calculates these L_{ASmx} values using the 'level' flight step in the procedure steps profile. The profiles themselves use the second highest departure weight; modelers using the Noise Impact Routing System (NIRS) determined this weight best represents average operational weights.
2) We collected actual spectra (not spectral class data) for each of the aircraft types at their maximum power settings. We used the manufacturer's INM submittal forms for the A320 and Embraer 145. We used BBN report 6039, "Revision of civil aircraft noise data for the Integrated Noise Model (INM)," for the Boeing 737-300 and the MD-83. We used Boeing departure spectral data for the 737-700 and the 777-200. The maximum power settings best represent the power setting where jet noise contributes the most to the low frequency components.
3) For each of the raw spectra, we calculated the L_{ASmx} values. We then applied a constant offset to each one-third octave Sound Pressure Level (SPL) so that the L_{ASmx} of the modified spectra matched the L_{ASmx} of the level overflights found in step 1.
4) For the five altitudes under consideration (20,000 to 40,000 feet MSL in 5,000 foot increments), we calculated the decibel offset provided by the simplified McAninch model equation. We added this offset to the L_{ASmx} values found in step 3.
5) For the modified spectra of step 3, we added a constant offset to each spectra to match the new L_{ASmx} associated with each altitude. We therefore have a spectra for each aircraft at each altitude; if logarithmically summed, each spectra will equal the associated L_{ASmx} value calculated in step 4.

Comparison of INM and McAninch L_{ASmx} values

Users can find the spectra calculated using the above process in a file accompanying this report. The tables below show the comparisons of the L_{ASmx} values calculated using the INM and those calculated using the McAninch model. The INM thrust (not necessarily the noise) increases as $10\log(1/\delta)$, a much smaller amount than the McAninch model. If the INM NPD data for a particular aircraft (and the thrust level of interest) increases the noise significantly with increasing thrust, than the INM and the McAninch model will match (e.g., MD-83). If the INM NPD data does not increase significantly with thrust, the noise predicted by the models will differ.

Note that these are L_{ASmx} values at 1000 feet from the source, and so have little contribution from the low frequencies. We expect the McAninch model to perform well where low frequencies dominate the received sound, such as occurs in long distance propagation. Note that in the tables below, the INM cannot calculate a noise level for some altitudes because its internal performance model breaks down. For these aircraft, the breakdown occurs from an inability to accelerate to the selected indicated airspeed using the available thrust at that altitude.

Table D-2. L_{ASmx} values at 1000 feet; INM

Aircraft types	Aircraft Altitude in feet				
	20,000	25,000	30,000	35,000	40,000
737-300	78.1	79.4	81.1	83.4	-
737-700	76.8	78.9	81.5	84.4	-
777-200	79.2	80.0	81.1	82.5	84.3
A320	76.5	78.3	80.7	83.7	-
EMB-145	72.4	73.9	75.9	-	-
MD-83	82.8	86.0	89.3	92.9	-

Table D-3. L_{ASmx} values at 1000 feet; McAninch Model

Aircraft types	Aircraft Altitude in feet				
	20,000	25,000	30,000	35,000	40,000
737-300	87.6	90.3	93.1	96.1	99.2
737-700	86.6	89.3	92.1	95.1	98.2
777-200	87.1	89.8	92.6	95.6	98.7
A320	85.1	87.8	90.6	93.6	96.7
EMB-145	77.1	79.8	82.6	85.6	88.7
MD-83	85.5	88.2	91.0	94.0	97.1

Table D-4. Difference in L_{ASmx} values at 1000 feet; INM minus McAninch Model

Aircraft types	Aircraft Altitude in feet				
	20,000	25,000	30,000	35,000	40,000
737-300	-9.5	-10.9	-12.0	-12.7	-
737-700	-9.8	-10.4	-10.6	-10.7	-
777-200	-7.9	-9.8	-11.5	-13.1	-14.4
A320	-8.6	-9.5	-9.9	-9.9	-
EMB-145	-4.7	-5.9	-6.7	-	-
MD-83	-2.7	-2.2	-1.7	-1.1	-

Appendix E: GCNP Sound Level Measurements of High Altitude Jet Aircraft

Introduction

In support of the joint US Department of Transportation (DOT), Federal Aviation Administration (FAA) and US DOI National Park Service (NPS) Alternative Dispute Resolution (ADR) process related to aircraft overflights of Grand Canyon National Park (GCNP), sound level measurements of high altitude jet aircraft were conducted in the vicinity of the GCNP North Rim. Specifically, measurements were conducted at Hancock Knoll (36••23' 43", -112••58' 07" - "HNK") and Swamp Point (36••20' 08", -112••20' 57"; - "SWP") between July 16 and 19, 2004. Figures E-1 through E-4 illustrate the location of the two measurement sites. HNK was located on a relatively flat plain at an altitude of approximately 5,900 feet MSL. The closest point on that canyon rim from HNK was almost 2 miles to the East. SWP was located on the rim about 30 miles to the East of HNK, at an altitude of approximately 7,500 feet MSL.

Approximately 16½ hours of simultaneous acoustic observer logs and sound level data were collected at HNK. Similarly, approximately 9 ½ hours of data were collected at SWP. A total of 18 hours of measurements were planned, however intermittent rain showers and some equipment problems interrupted measurements several times.

Measurement data included simultaneous acoustic observer logs and sound level data collected using the specialized Volpe Low Amplitude Recording Equipment (VOLARE). Additionally, separate, continuous sound level and wind speed and direction data were collected using the NoiseLogger™ system at HNK. Table E-1 presents a summary of the measurement data collected at HNK. Simultaneous sound level measurements on the ground and at five feet above the ground were also conducted at SWP. Table E-2 presents a summary of the measurement data collected at SWP.

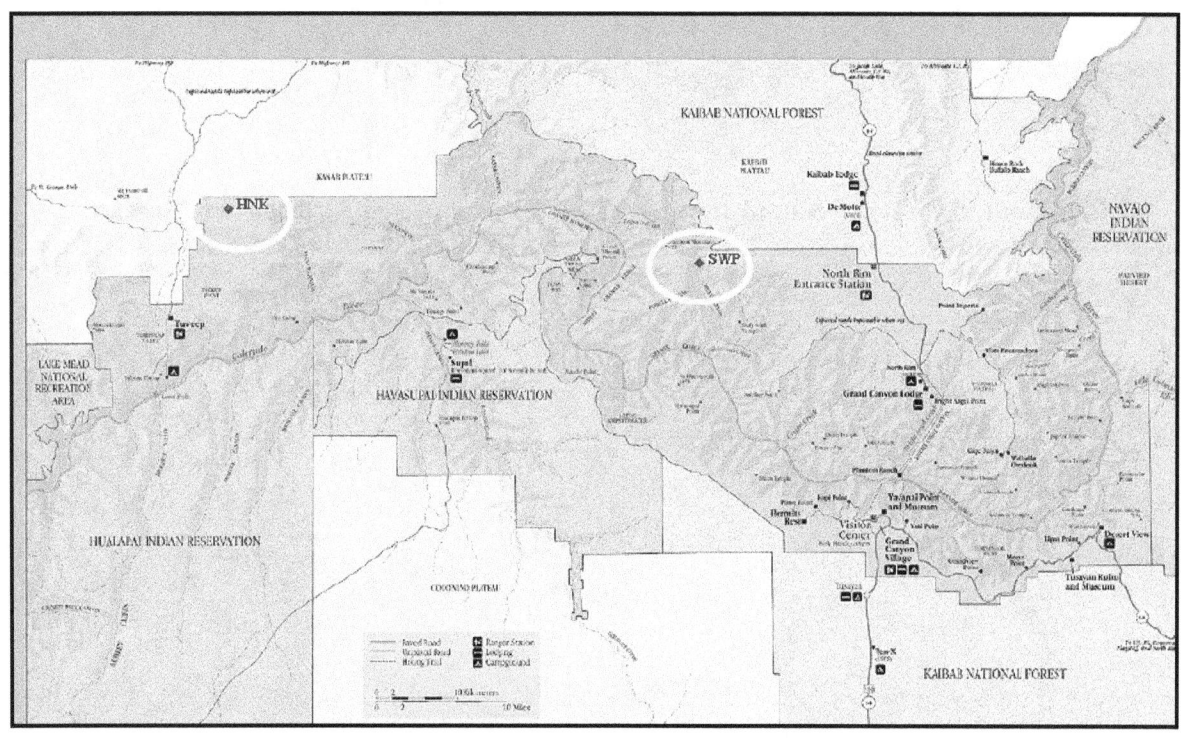

Figure E-1. Relative Location of Measurement Sites in Grand Canyon

Figure E-2 presents the terrain elevation profile between HNK and SWP.

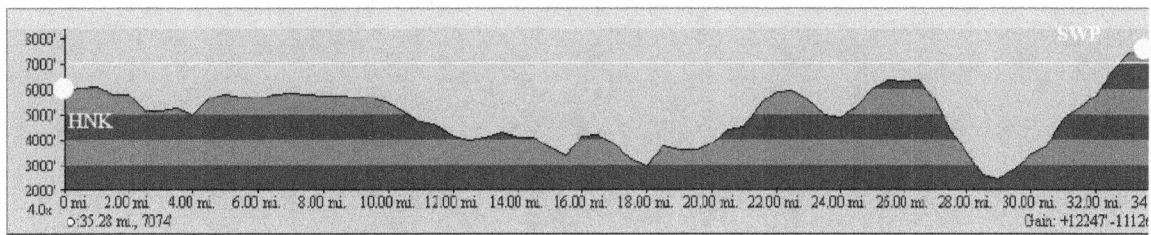

Figure E-2. Elevation Profile: Hancock Knoll to Swamp Point

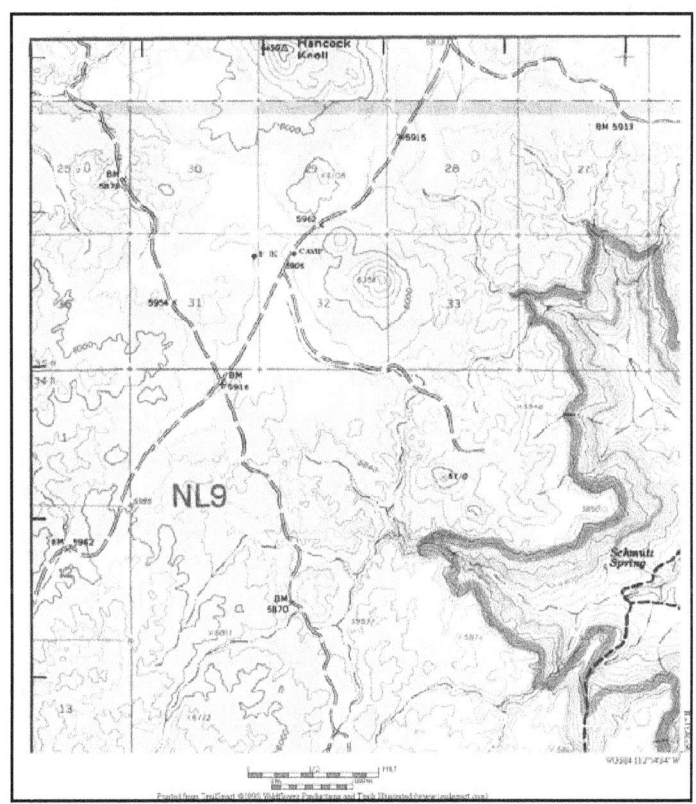

Figure E-3. Hancock Knoll Measurement and Camp Sites

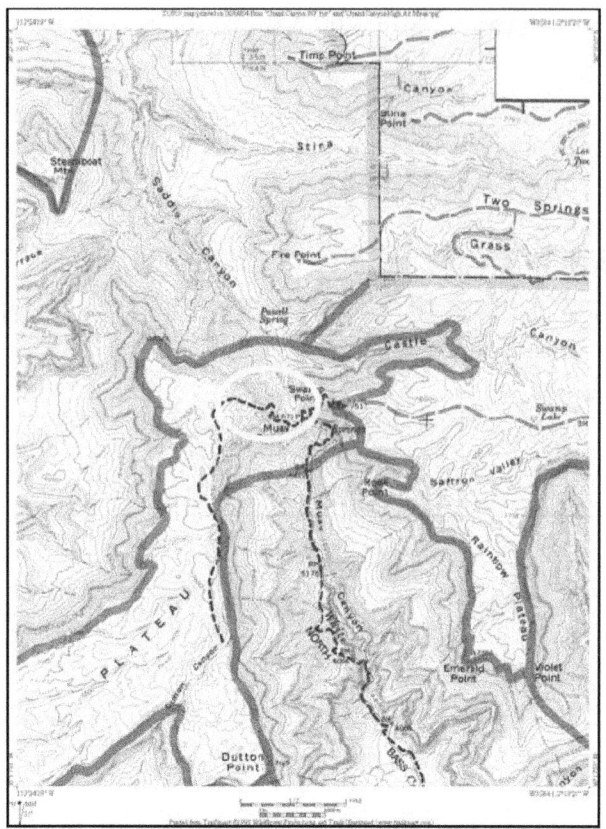

Figure E-4. Swamp Point Measurement Site

Figure E-5 is a photograph of some of the instrumentation located at HNK. In the foreground are the solar panels and electronics for the NoiseLogger™ system. Further back, from left to right, are the NoiseLogger™ anemometer, microphone/tripod/windscreen, and VOLARE microphone/tripod with two-stage windscreen. Figure E-6 is a photograph of the VOLARE system electronics, which were located about 100 feet from the microphones.

Figure E-5. Hancock Knoll Instrumentation, Part 1

Figure E-6. Hancock Knoll Instrumentation, Part 2

Figure E-7 presents the instrumentation located at the SWP.

Figure E-7. Swamp Point Measurement Site

Table E-1. Hancock Knoll Measurement Summary Statistics

	Jets	Props	All Aircraft
Events (#)	158	19	177
Time Audible (hh:mm:ss)	6:03:25	0:38:39	6:42:04
Overall Percent Time Audible (%)	37	4	41
Percent Time Audible 7/16: 22:00 – 23:30 (%)	27	0	27
Percent Time Audible 7/17: 12:00 – 15:00 (%)	36	3	39
Percent Time Audible 7/17: 16:00 – 19:00 (%)	39	3	42
Percent Time Audible 7/18: 07:00 – 10:00 (%)	40	11	51
Percent Time Audible 7/18: 12:00 – 15:00 (%)	33	4	37
Percent Time Audible 7/19: 07:00 – 10:00 (%)	53	2	55
"Single Events"[7] (#)	80	1	81
Average "Min Rise/Fall"[8] (dBA)	17 (104)	16 (5)	16 (109)
Minimum "Min Rise/Fall" (dBA)	2	7	2
Maximum "Min Rise/Fall" (dBA)	37	26	37

---

[7] "Single Events" are defined as acoustic states designated as aircraft which were bounded before and after by natural sounds (i.e., not other aircraft).

[8] "Min Rise/Fall" is defined as the minimum of either the sound level rise or fall, relative to L_{ASmx}, from the beginning or end of the event, respectively. The number in parentheses represents the total number of events used to calculate Min Rise/Fall.

Table E-2. Swamp Point Measurement Summary Statistics[9]

	Jets	Props	All Aircraft
Events (#)	96	11	107
Time Audible (hh:mm:ss)	2:43:39	0:23:53	3:07:32
Overall Percent Time Audible (%)	14	2	16
"Single Events" (#)	23	1	24
Average "Min Rise/Fall" (dBA)	13(23)	5 (1)	13 (24)
Minimum "Min Rise/Fall" (dBA)	6.3	5	5
Maximum "Min Rise/Fall" (dBA)	29.3	5	29.3

Example Time History

Figure E-8 below presents an example time history. Included on the graphic are time histories of both the A-weighted, slow-scale sound level and the acoustic state, as determined by the observer on-site during the measurements. Note that excellent signal-to-noise ratios are illustrated for several events. The individual event sound level time histories measured are typically asymmetrical about the maximum sound level; rather, as theory predicts, there is a brief, rapid rise in sound levels followed by more gradual drop off over time, associated with low-frequency jet noise after the aircraft has passed overhead.

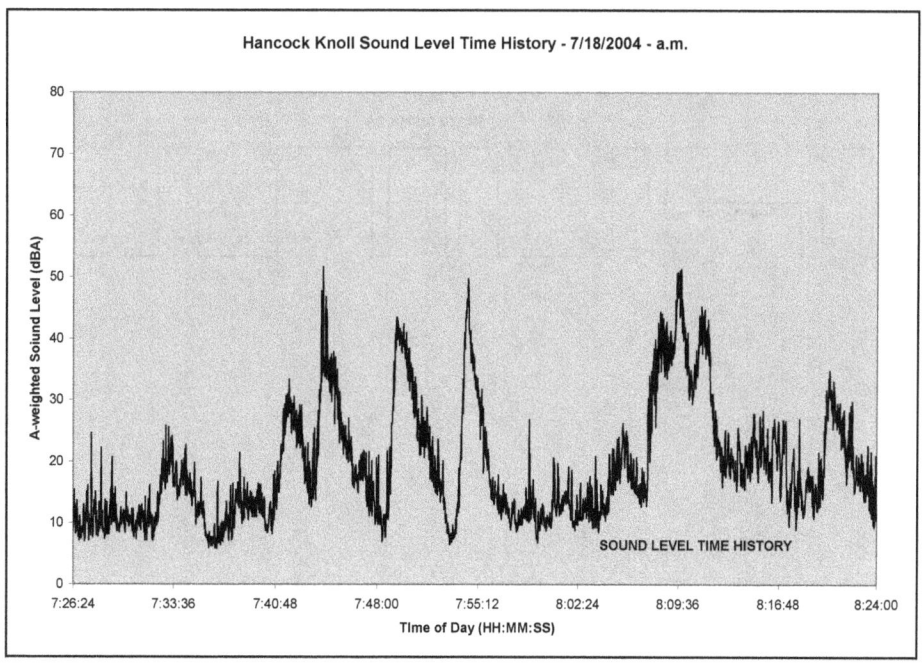

Figure E-8. Example Time Histories Measured at Hancock Knoll

---

[9] These data include audibility logging for the entire day on 7/18, including periods when there were no acoustic measurements taken. This, and some equipment problems, accounts for the high number of events but the low number of 'Single' events.

Figure E-9 presents representative aircraft spectra at time of maximum A-weighted sound level (L_{ASmx}). Aircraft identified as Airbus A320 during the initial ETMS data mining are presented in the figure. Note that the slant distance at time L_{ASmx} was emitted for the individual events ranges from approximately 5 to 14 nautical miles.

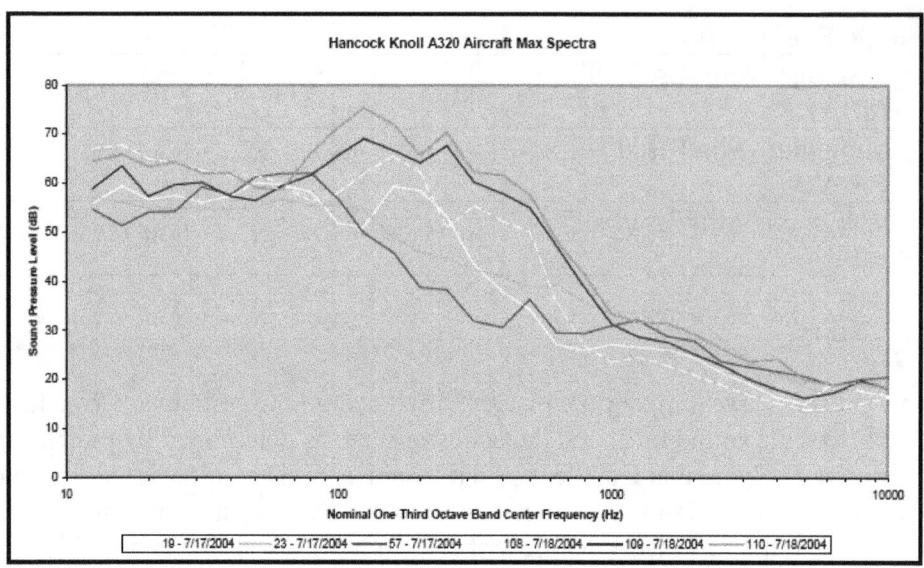

Figure E-9. Representative A320 Aircraft Maximum Spectra Measured at Hancock Knoll

Sound Level Histograms

The following graphics present histograms of the sound level data collected at both HNK and SWP. The graphics include all sounds (i.e., both aircraft as well as the sounds of nature) during the measurement periods.

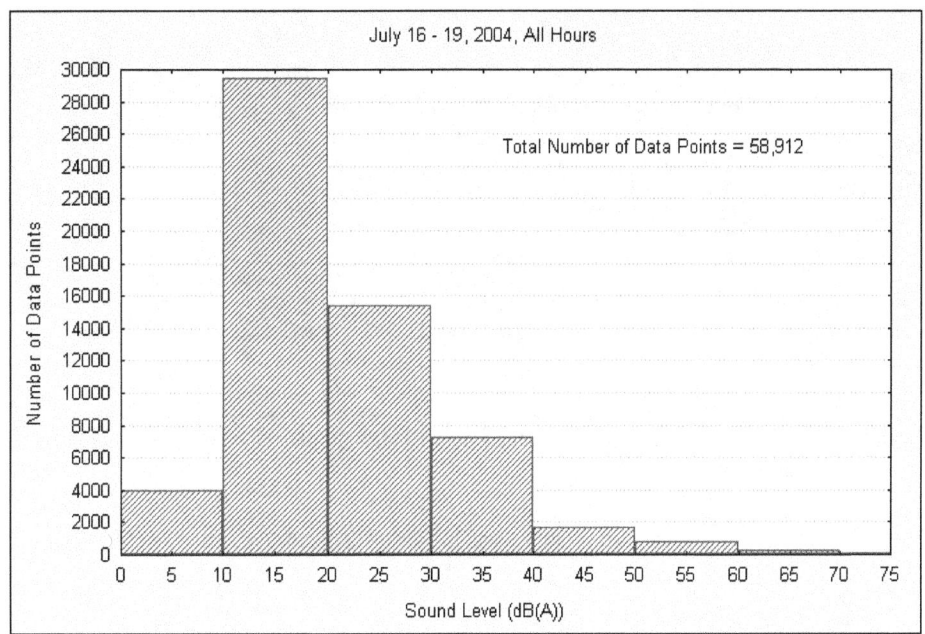

Figure E-10. Hancock Knoll Sound Level Histogram – All Data

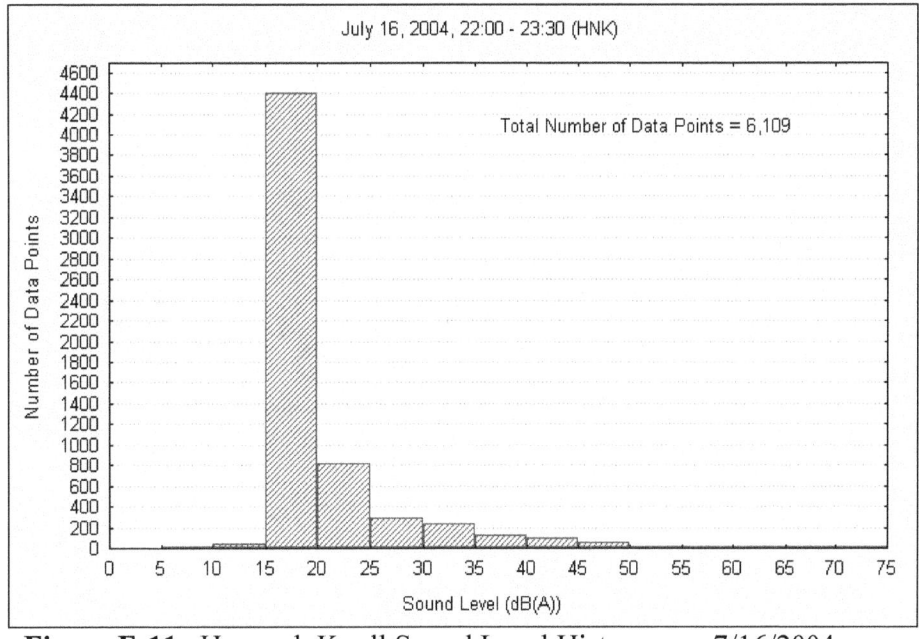

Figure E-11. Hancock Knoll Sound Level Histogram – 7/16/2004, p.m.

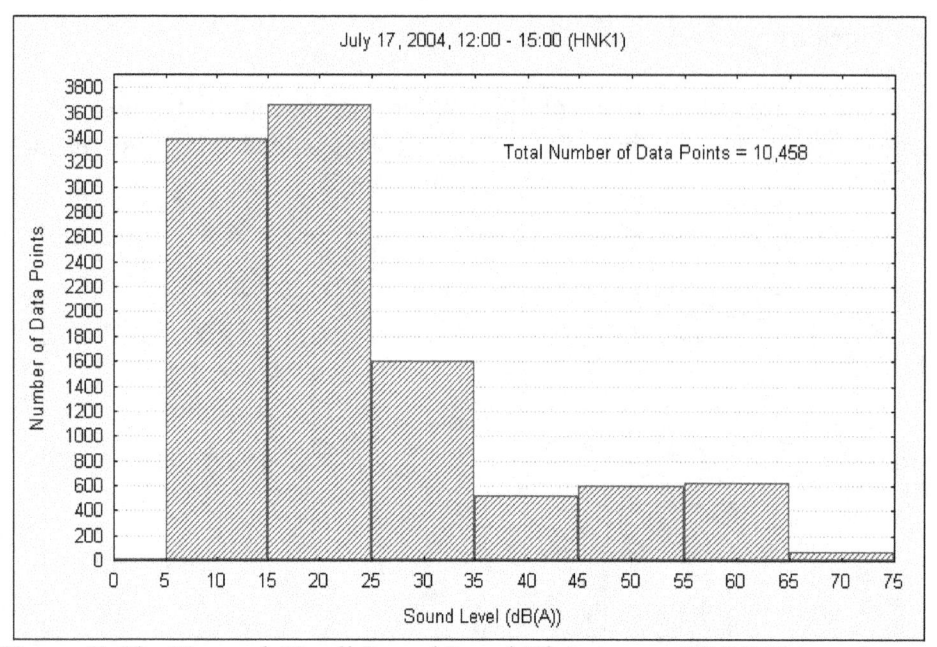

Figure E-12. Hancock Knoll Sound Level Histogram – 7/17/2004, early p.m.

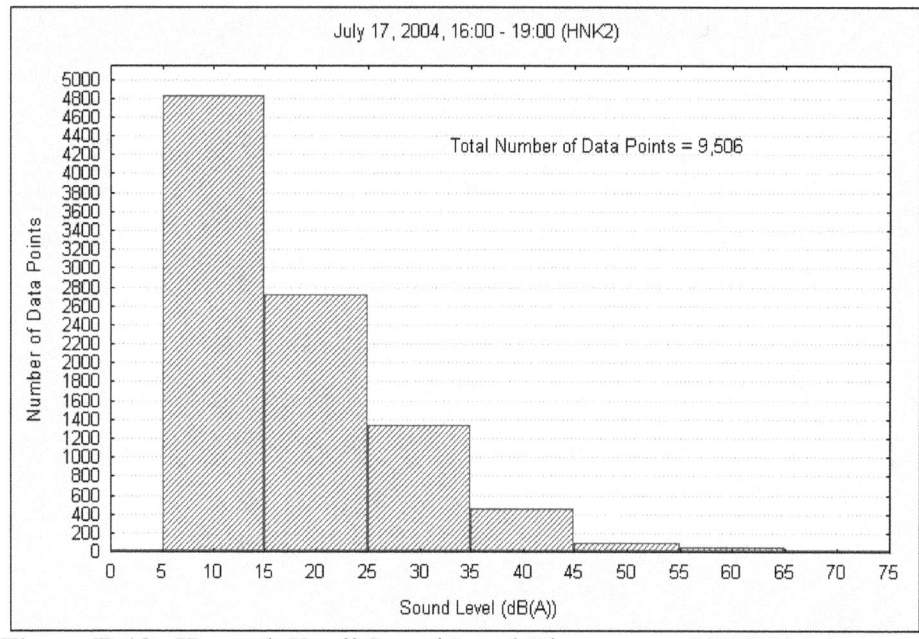

Figure E-13. Hancock Knoll Sound Level Histogram – 7/17/2004, late p.m.

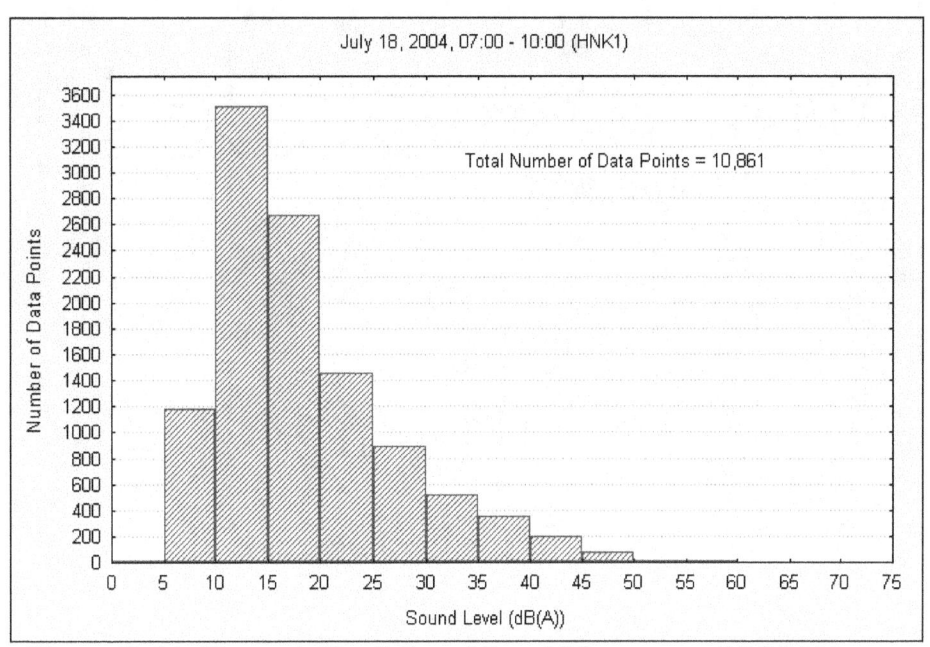

Figure E-14. Hancock Knoll Sound Level Histogram – 7/18/2004, a.m.

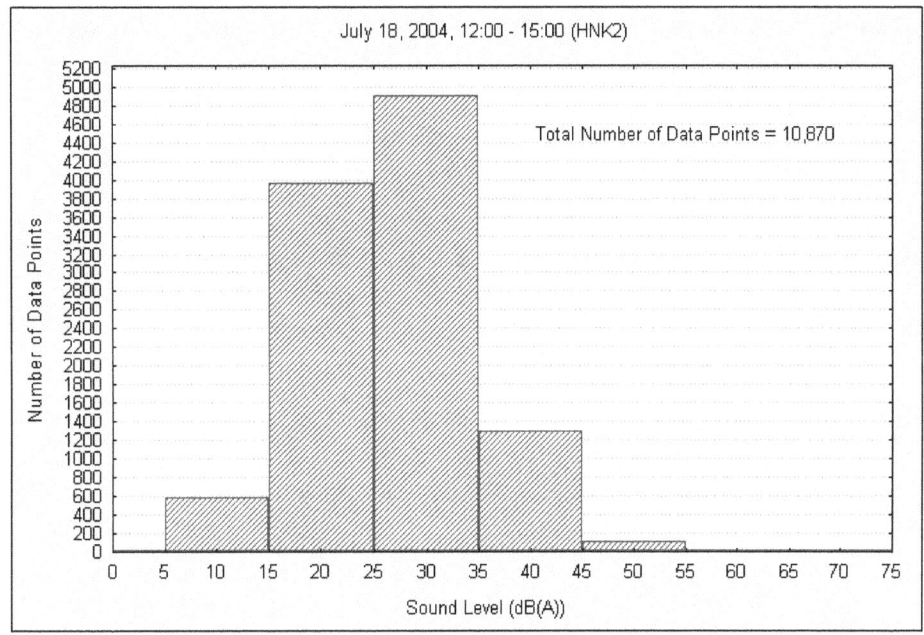

Figure E-15. Hancock Knoll Sound Level Histogram – 7/18/2004, p.m.

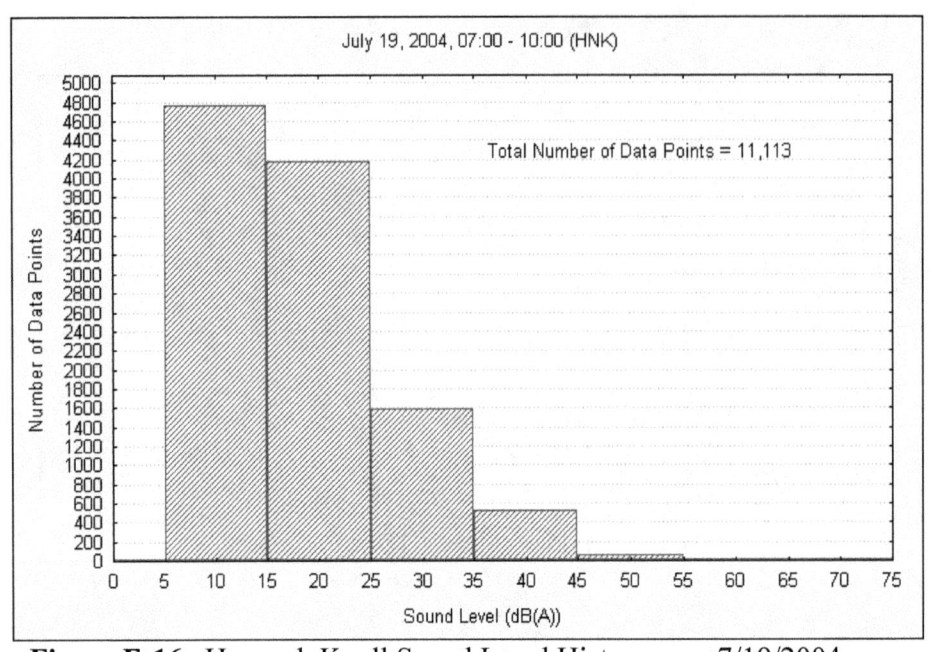

Figure E-16. Hancock Knoll Sound Level Histogram – 7/19/2004, a.m.

The following graphics present similar histograms of the sound level data collected at Swamp Point.

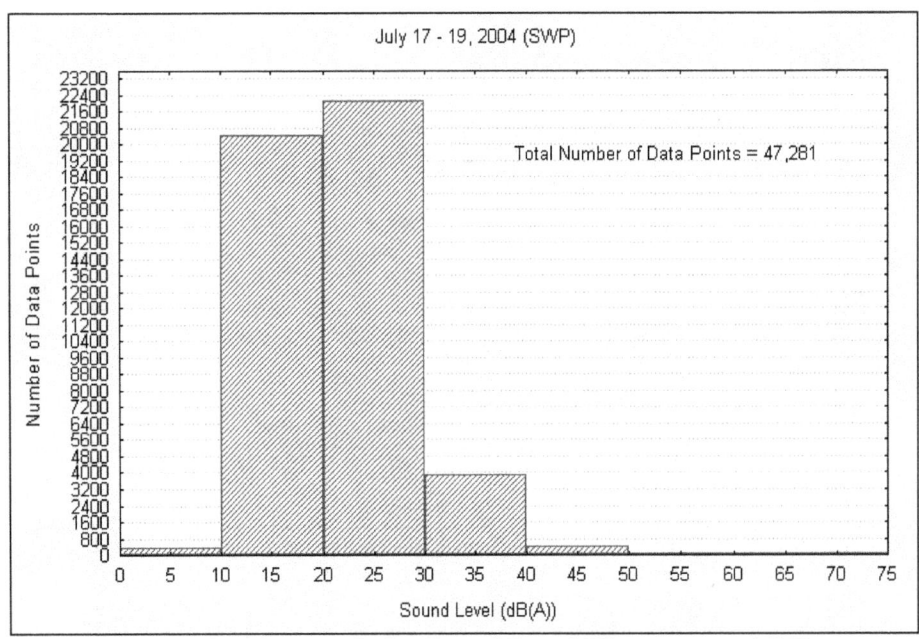

Figure E-17. Swamp Point Sound Level Histogram – All Data

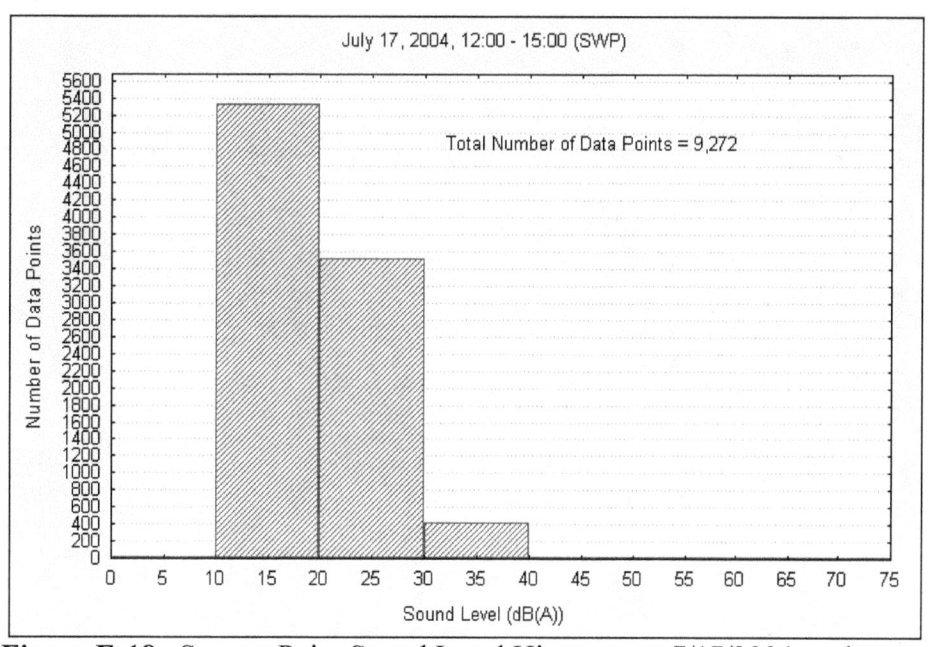

Figure E-18. Swamp Point Sound Level Histogram – 7/17/2004, early p.m.

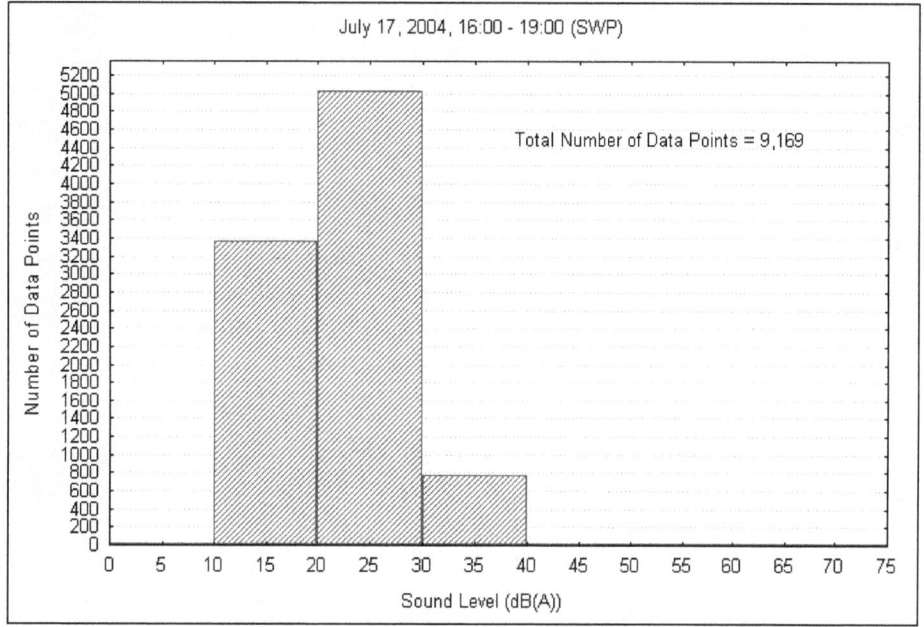

Figure E-19. Swamp Point Sound Level Histogram – 7/17/2004, late p.m.

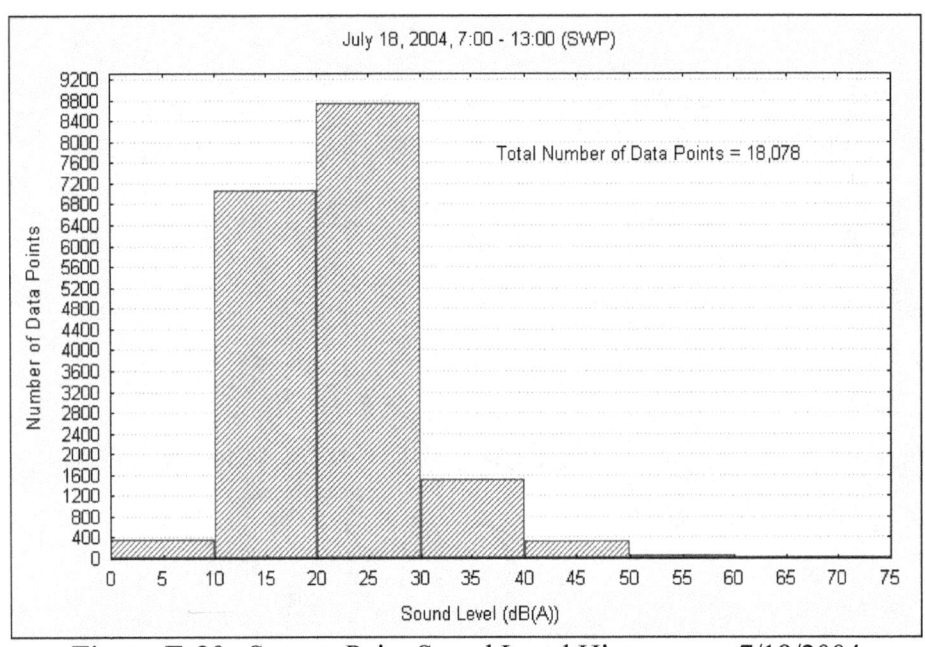

Figure E-20. Swamp Point Sound Level Histogram – 7/18/2004

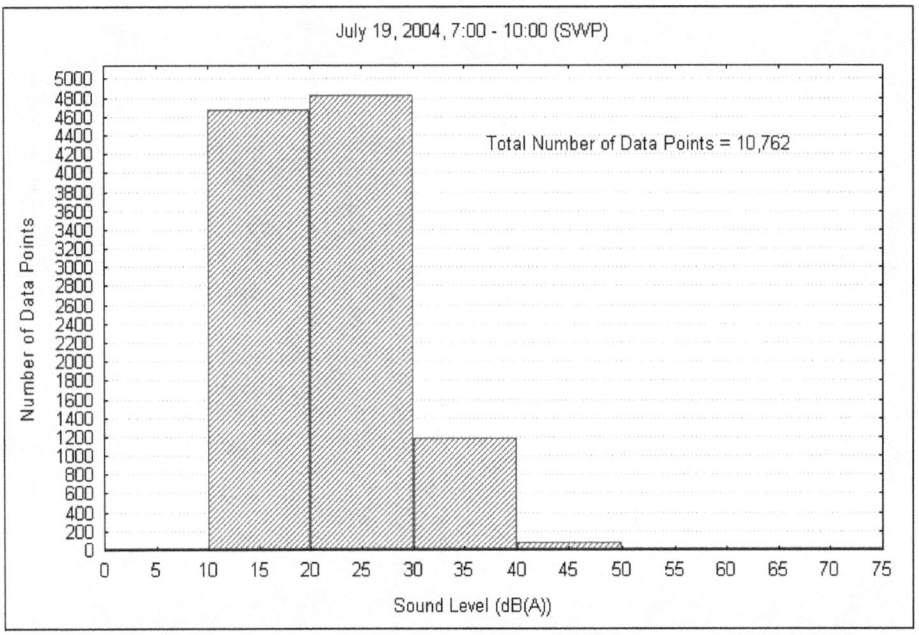

Figure E-21. Swamp Point Sound Level Histogram – 7/19/2004

Natural Sounds Time Period Histograms

Figures E-22 through E-32 present histograms of the natural ambient segment durations at both HNK and SWP. Time intervals for all data are one minute (i.e., the first bar represents time periods less than or equal to one minute, the second bar represents time periods greater than one minute and less than or equal to two minutes, etc.)

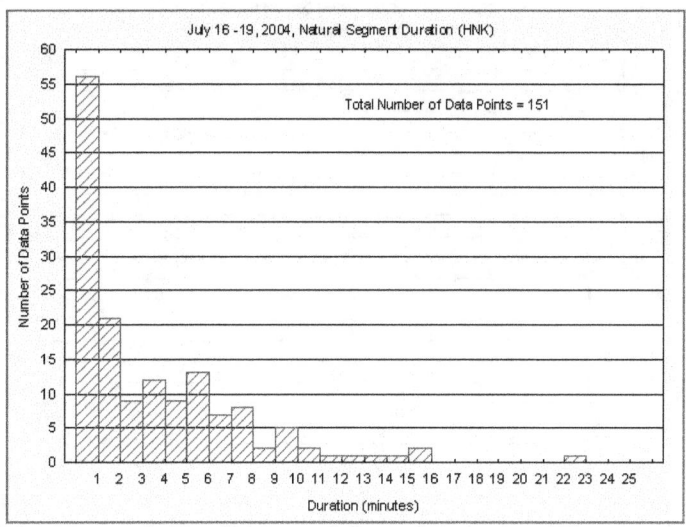

Figure E-22. Hancock Knoll Natural Sound Duration Histogram – All Data

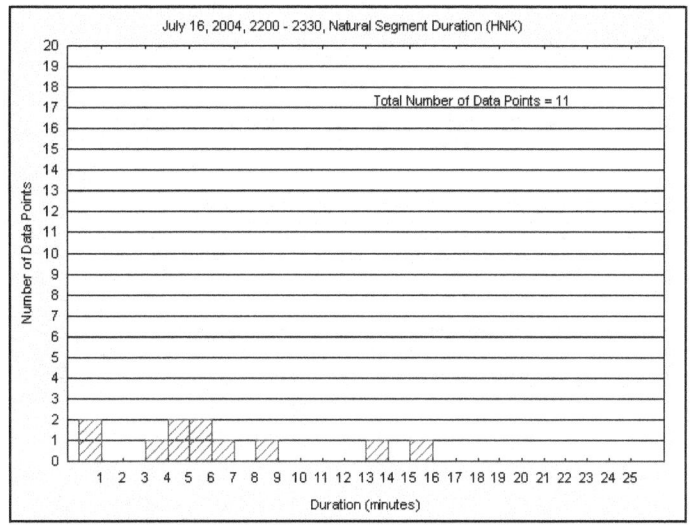

Figure E-23. Hancock Knoll Natural Sound Duration Histogram – 7/16/2004

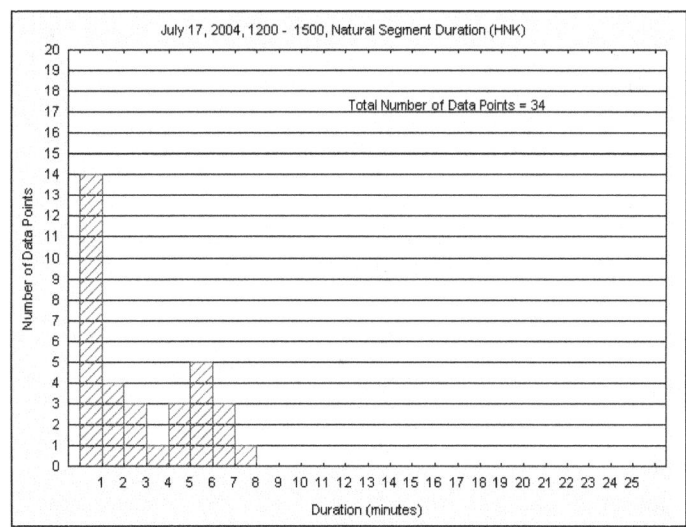

Figure E-24. Hancock Knoll Natural Sound Duration Histogram – 7/17/2004, early p.m.

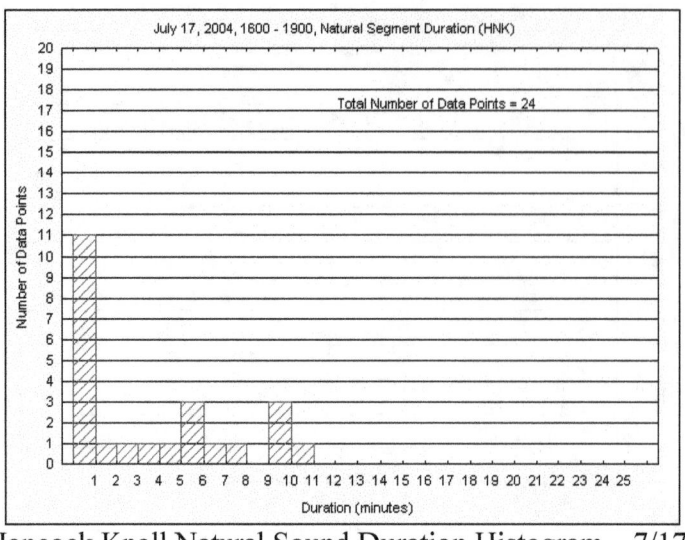

Figure E-25. Hancock Knoll Natural Sound Duration Histogram – 7/17/2004, late p.m.

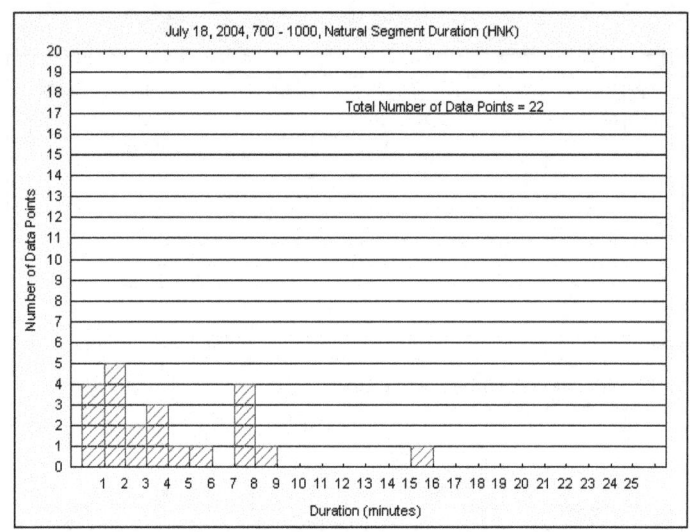

Figure E-26. Hancock Knoll Natural Sound Duration Histogram – 7/18/2004, a.m.

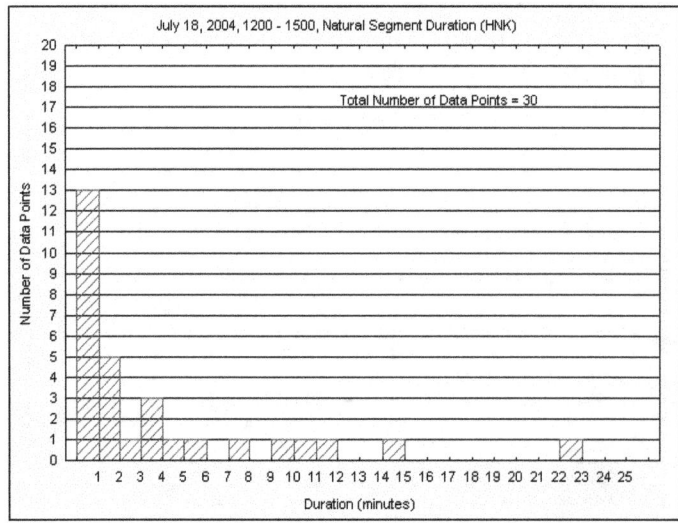

Figure E-27. Hancock Knoll Natural Sound Duration Histogram – 7/18/2004, p.m.

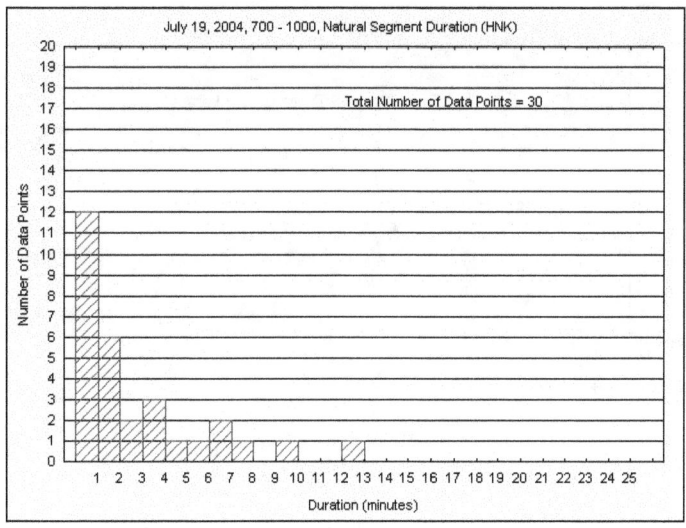

Figure E-28. Hancock Knoll Natural Sound Duration Histogram – 7/19/2004

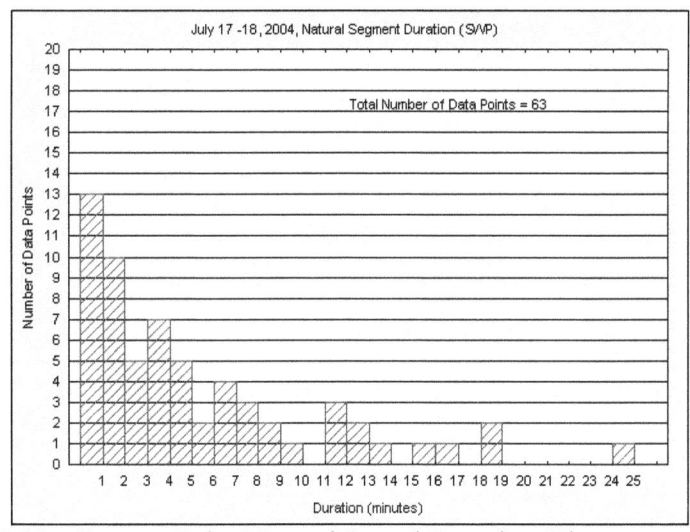

Figure E-29. Swamp Point Natural Sound Duration Histogram – All Data

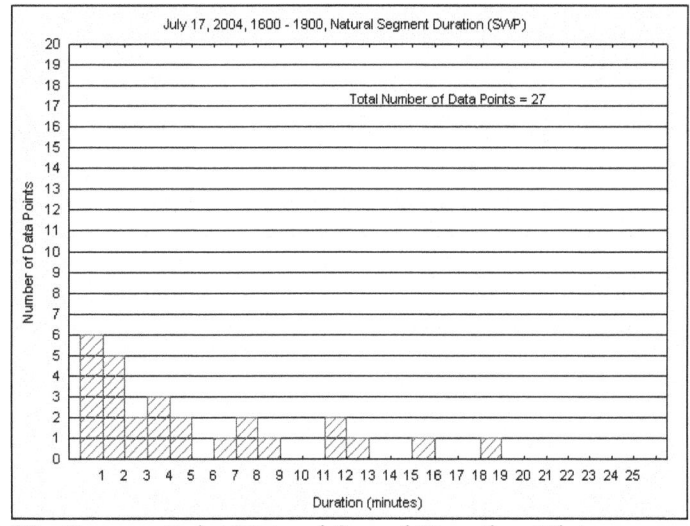

Figure E-30. Swamp Point Natural Sound Duration Histogram – 7/17/2004

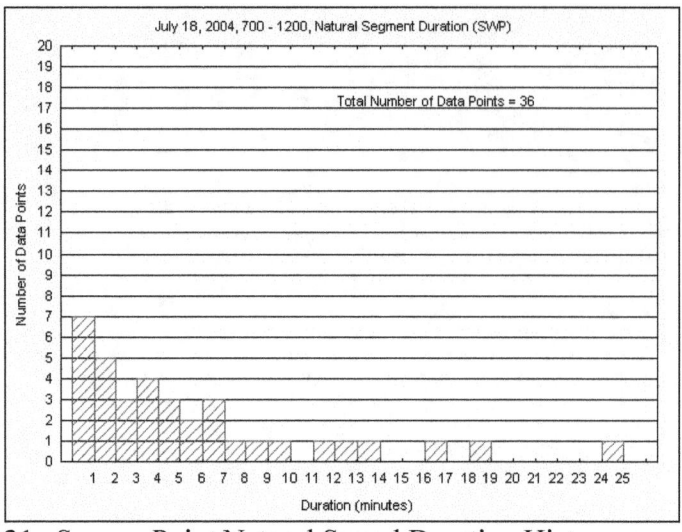

Figure E-31. Swamp Point Natural Sound Duration Histogram – 7/18/2004

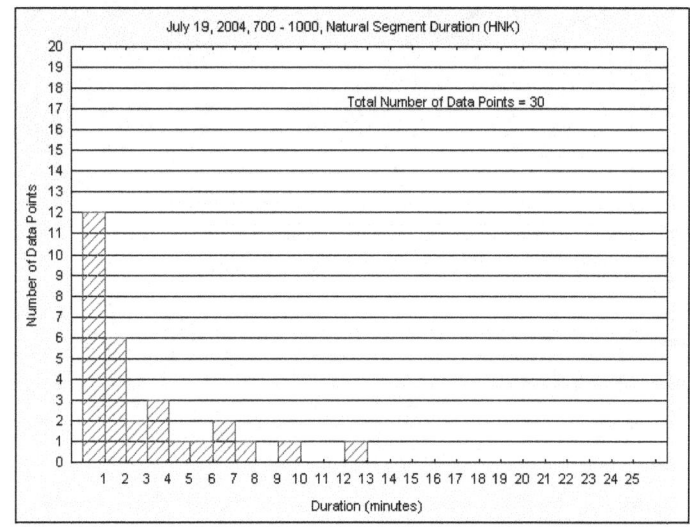

Figure E-32. Swamp Point Natural Sound Duration Histogram – 7/19/2004

Comparative Acoustic State Logs

During some measurement periods at HNK concurrent acoustic state logs were maintained by two personnel. Figure E-33 presents an overlay of the acoustic states assigned by the two people during one such measurement period. Note that in general there is excellent agreement between the two logs. There are subtle differences in start and end times between the two logs, as well as a couple of identified events which differed, however these differences result in less than seven percent overall differences in jet acoustic state audibility times, and less than four percent difference for props and natural.

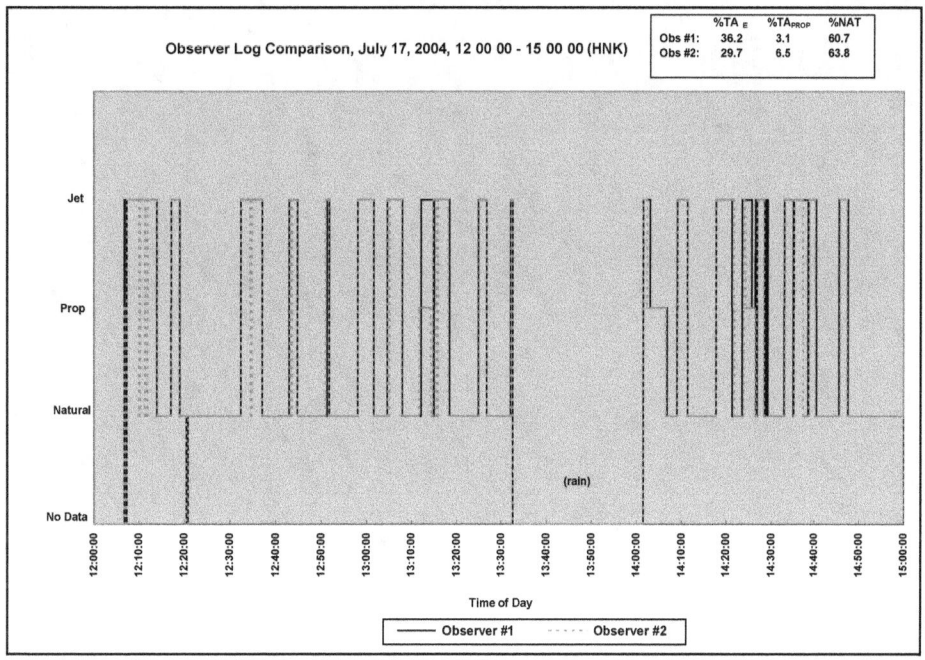

Figure E-33. Summary of Concurrent Acoustic States

E-20

Appendix F: Statistical Definitions

Appendix F presents both the descriptions and formulae for the statistical measures utilized in this document.

Table F-1. Definitions of Statistical Measures

Name	Description	Formula
Overall Error	Average of the squared differences between the modeled and measured data	$\text{Error}_{\text{Overall}} = \sqrt{\dfrac{\sum(S_{\text{modeled}} - S_{\text{measured}})^2}{N}}$
Bias	Average of the differences between the modeled and measured data	$\text{Bias} = \dfrac{\sum(S_{\text{modeled}} - S_{\text{measured}})}{N}$
95% Confidence Interval	The ± value around the bias for the 95% confidence interval	$\text{C.I.} = 1.96 * \sqrt{\dfrac{\text{Var}_{\text{Delta}}}{N}}$ where $\text{Var}_{\text{Delta}} = \dfrac{\sum[(S_{\text{modeled}} - S_{\text{measured}}) - \overline{(S_{\text{modeled}} - S_{\text{measured}})}]^2}{N-1}$
Random Error	Standard deviation of the residuals from the logistic regression for modeled vs. measured data.	$\text{Error}_{\text{random}} = \sqrt{\dfrac{\sum(S_{\text{residual}} - \overline{S}_{\text{residual}})^2}{N-1}}$ where $S_{\text{residual}} = S_{\text{measured}} - S_{\text{measured(pred)}}$ and $S_{\text{measured(pred)}} = \dfrac{b_0}{1 + b_1 * e^{(b_2 * S_{\text{modeled}})}}$ where b_1, b_2, and b_3 are calculated by Statistica
Correlation Coefficient	Correlation coefficient between the modeled and measured data	$\text{C.C.} = \dfrac{\sum[(S_{\text{modeled}} - \overline{S}_{\text{modeled}}) * (S_{\text{measured}} - \overline{S}_{\text{measured}})]}{\sqrt{\sum(S_{\text{modeled}} - \overline{S}_{\text{modeled}})^2 * \sum(S_{\text{measured}} - \overline{S}_{\text{measured}})^2}}$

where:

N = Number of data points

S_{modeled} or S_{measured} = Individual data points

$\overline{S}_{\text{modeled}}$ or $\overline{S}_{\text{measured}}$ = Mean values

Appendix G: Development of Reference Noise Data for High Altitude Jets

In order to approximate an appropriate thrust setting for use in modeling high altitude jet flyovers in this effort, an analytical method was developed to use acoustic data, paired with ETMS flight tracking data (highlighted in Appendix C), to select an appropriate thrust value from the available INM NPD curves.

First, measured data points (measurements summarized in Appendix E) for all like aircraft were grouped with the aircraft-specific INM NPD curves. Next, the assumption was made that the NPD curves behave in a log-linear manner beyond the distances included in the data. This assumption allowed a first-order regression fit to be calculated based on the curves and ETMS flight tracking data. The NPD curve that came closest to the regression line was assumed to be the appropriate power setting. As with any first-order regression fit, two values needed were calculated: slope of the regression and its intercept.

The regression slope was calculated based solely on the NPD curves. It was assumed that all NPD values (for the same aircraft type at distances greater than 10,000 feet) have a log-linear relationship of approximately the same slope. This assumption results in the slope of a regression being independent of thrust setting. Accordingly, the slope could be approximated without making use of the slant range associated with the ETMS flight tracking data. To aggregate all the NPD curves for a particular aircraft, a slope was calculated separately for each thrust setting. Then, the slopes of all thrust settings were averaged for each aircraft to approximate an overall slope. This slope would serve as the regression line slope.

While the slope of regression was assumed to not vary based on thrust setting, deriving an intercept required the use of the individual measurement point's SEL and associated slant range from the ETMS flight tracking data. For each aircraft type, a regression analysis was performed using the flight tracking data points, solving for the intercept. The final regression was plotted against the INM NPD curves. The curve that came closest to the regression equation was selected as the thrust setting. The resultant plots are below, including the INM NPD data, individual measurement points, calculated regression lines and a 95% confidence intervals around the estimations.

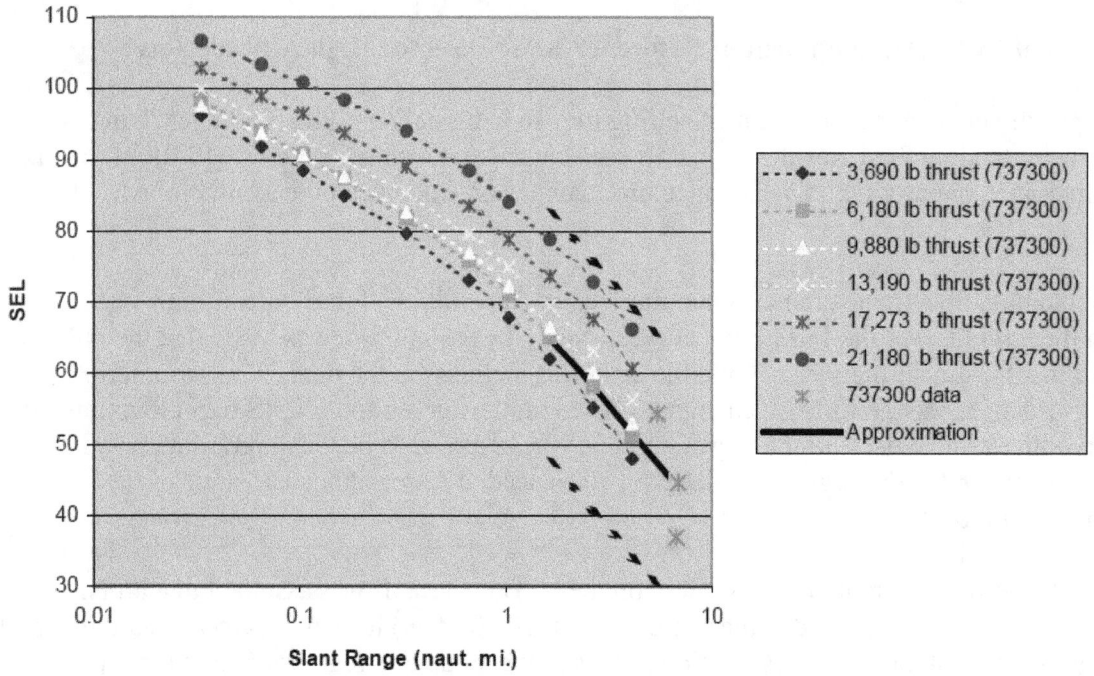

SEL vs Slant Range for 737300

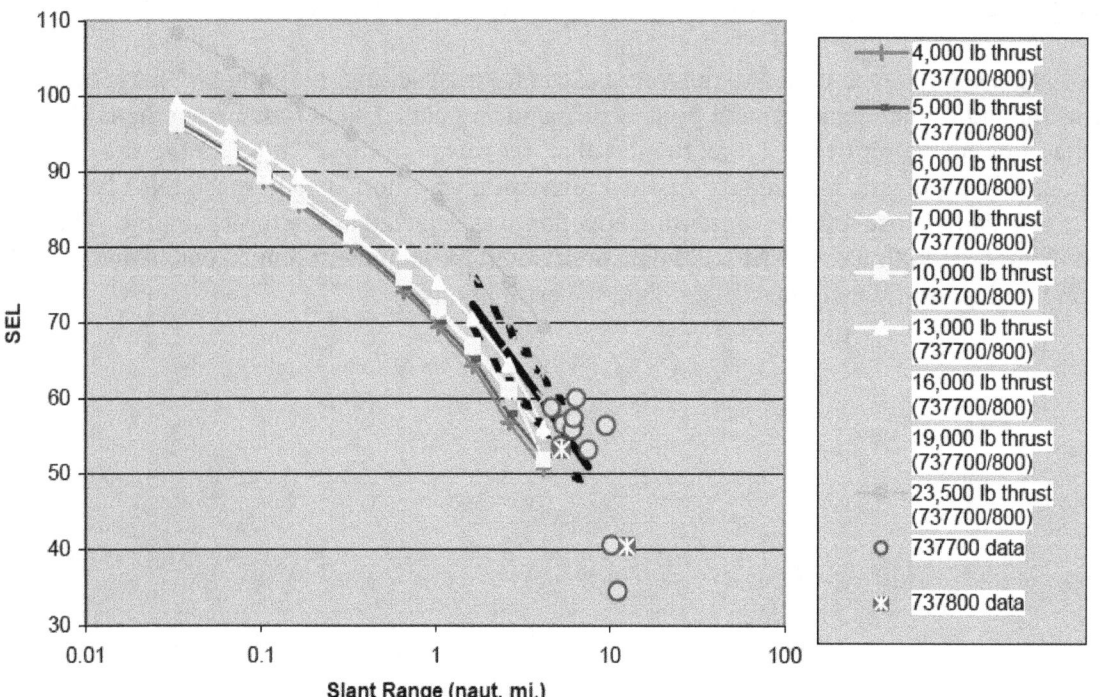

SEL vs Slant Range for 737700/800

G-2

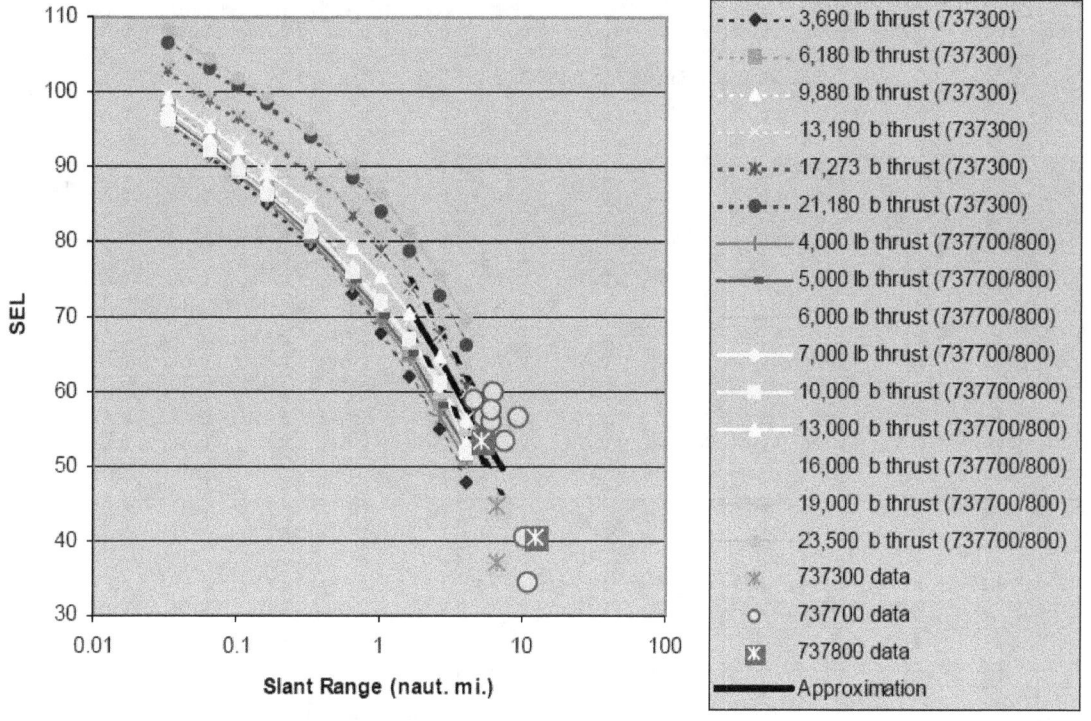

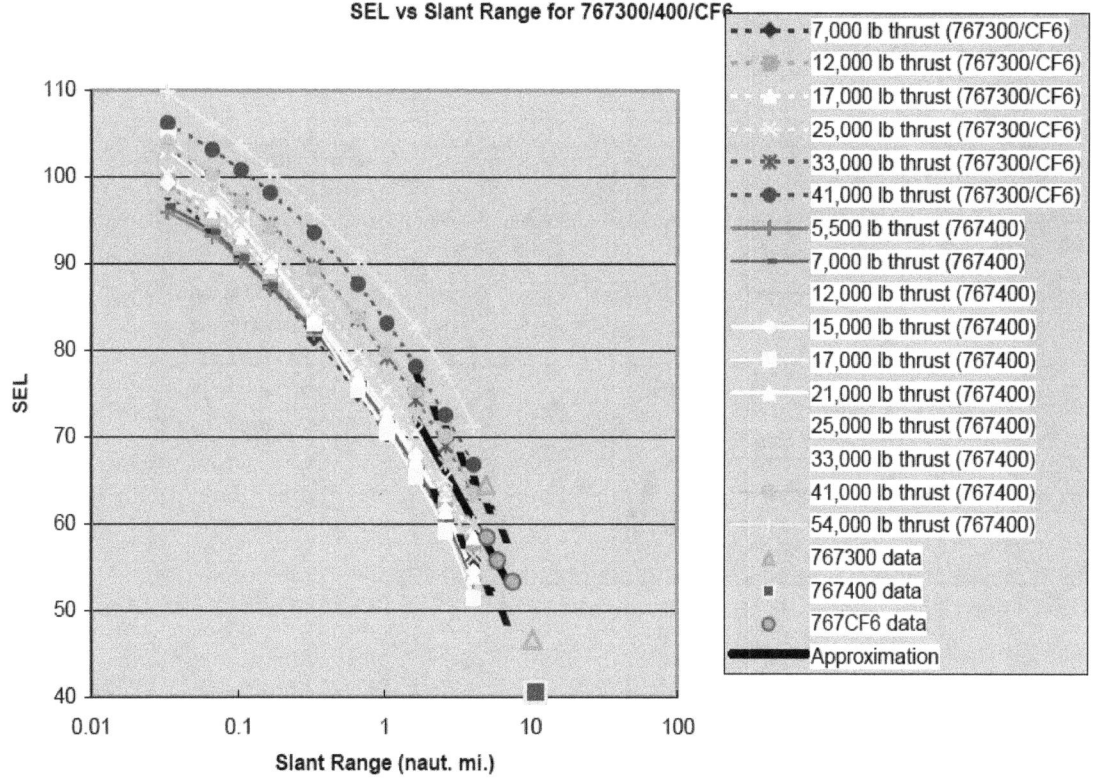

G-3

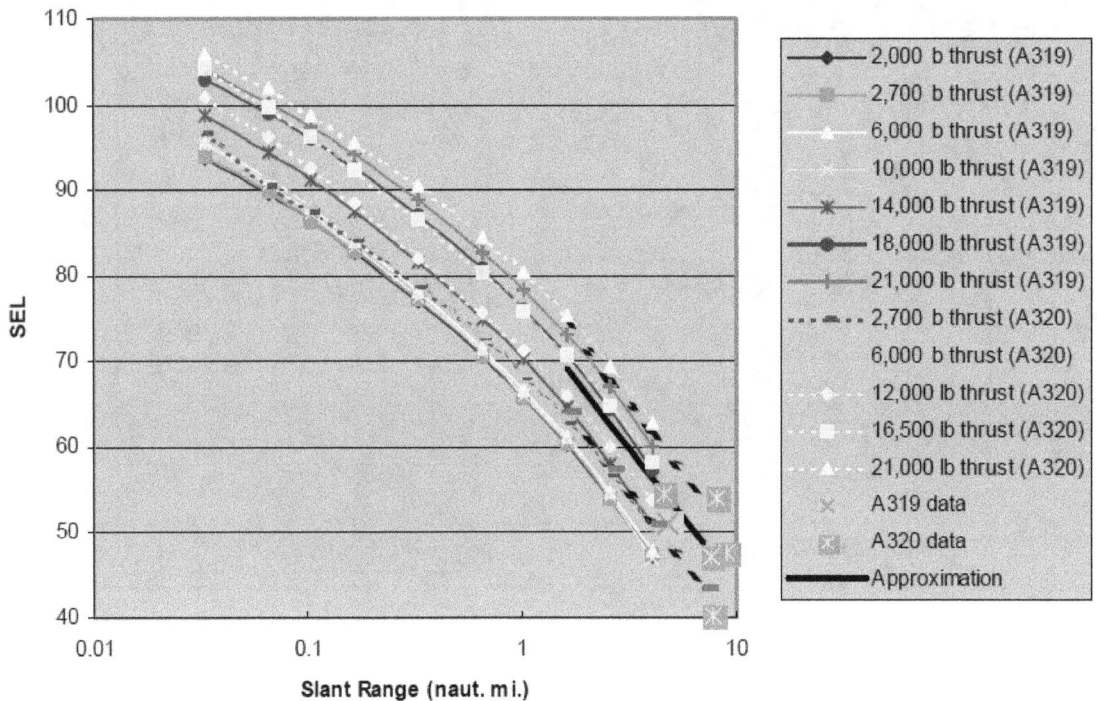
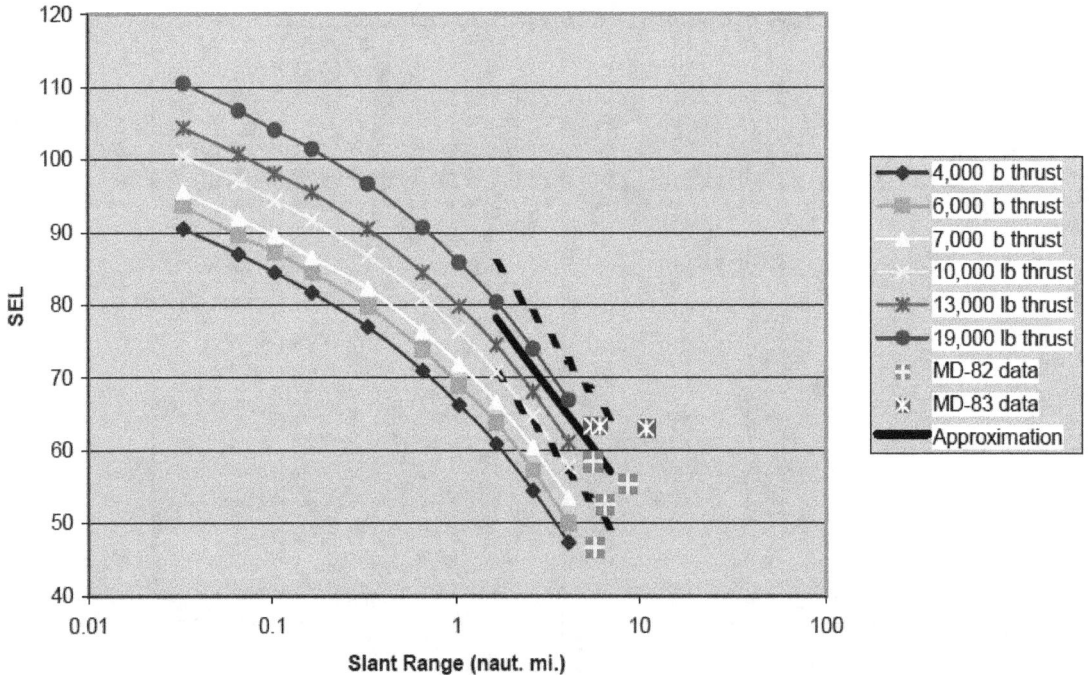

G-4

Appendix H: Audibility Calculations for National Parks

During the GCNP MVS it was agreed that audibility would be computed using:

$$d' = eta*sqrt(bw)*s/n \qquad (1)$$

where, *eta* = listener efficiency, *bw* = bandwidth, *s* = signal, and *n* = noise. N should consist of amb+easn, where amb = ambient sound and easn = equivalent auditory system noise. EASN is computed from Equation (1) by using threshold of hearing (earspc) for s, 0 for amb, and setting d' = 1.5. (for simplicity, s, n, amb, earspc, are easn are intensity, not dB).

All modelers were provided measured ambient spectra to which easn had been added, i.e., the file amb_spc2.csv consisted of amb+easn. EASN was based on earspc from ISO R226-1961.

NMSim MVS calculations were done with Equation (1), using the supplied amb+easn as n. Subsequent NMSim calculations have been done with Equation (1), but using ambient-only "amb" instead of amb+easn, as n. Current NMSim d' is thus computed using n = amb.

INM MVS calculations were done with Equation (1), using the supplied amb+easn as n. Subsequent INM calculations have been done with Equation (1), using the supplied amb as amb, and adding to that the quantity of earspc. Current INM d' is thus computed using n = amb+earspc, with earspc based on ISO 389-7:1998.

For future modeling (including FICAN re-analysis of MVS past the October 2004 FICAN meeting), INM and NMSim will use Equation (1) with n = amb+easn, where:

amb = ambient alone. (For MVS analysis, easn will be subtracted from amb_spc2.csv.); and
easn is to be computed from Equation (1) using an appropriate earspc from ISO 389-1:1998.

Summary Table

	INM	**NMSim**
1999 MVS	Eq 1 & amb+EASN	Eq 1 & amb+EASN
2004 MVS	Eq 1 & amb+EASN+earspc*	Eq 1 & amb+EASN**
2004 Contours	Eq 1 & amb (w/o EASN)+earspc*	Eq 1 & amb (w/o EASN) **
FICAN Technical Report and any Future GCNP Analyses	Eq 1 & amb (w/ EASN taken into account in code, not amb, and earspc based on ISO 1998 (free field))	

*With earspc based on ISO 1998 (diffuse field).
**With earspc based on ISO 1961 (free field).